Advances in Combinatorial Optimization

Linear Programming Formulations of the
Traveling Salesman and Other Hard
Combinatorial Optimization Problems

Advances in Combinatorial Optimization

Linear Programming Formulations of the Traveling Salesman and Other Hard Combinatorial Optimization Problems

Moustapha Diaby
University of Connecticut, USA

Mark H Karwan
University at Buffalo,
The State University of New York, USA

World Scientific

NEW JERSEY · LONDON · SINGAPORE · BEIJING · SHANGHAI · HONG KONG · TAIPEI · CHENNAI · TOKYO

Published by

World Scientific Publishing Co. Pte. Ltd.
5 Toh Tuck Link, Singapore 596224
USA office: 27 Warren Street, Suite 401-402, Hackensack, NJ 07601
UK office: 57 Shelton Street, Covent Garden, London WC2H 9HE

Library of Congress Cataloging-in-Publication Data
Diaby, Moustapha.
    Advances in combinatorial optimization : linear programming formulations of the
traveling salesman and other hard combinatorial optimization problems / Moustapha Diaby
(University of Connecticut, USA), Mark H. Karwan (University at Buffalo, The State University
of New York, USA).
        pages cm
    Includes bibliographical references.
    ISBN 978-9814704878 (hardback : alk. paper)
    1. Combinatorial optimization. 2. Mathematical optimization. I. Karwan, Mark H., 1951–
II. Title.
    QA402.5.D524 2015
    519.6'4--dc23

                                                    2015026035

British Library Cataloguing-in-Publication Data
A catalogue record for this book is available from the British Library.

In-house Editors: Amanda Yun/Dipasri Sardar

Typeset by Stallion Press
Email: enquiries@stallionpress.com

Printed in Singapore

"*Hâtez-vous lentement, et sans perdre courage,*
Vingt fois sur le métier remettez votre ouvrage,
Polissez-le sans cesse, et le repolissez,
Ajoutez quelquefois, et souvent effacez."

Nicolas Boileau-Despréaux

Contents

About the Authors

Moustapha Diaby is Associate Professor of Production and Operations Management at the University of Connecticut. He received a PhD degree in Management Science/Operations Research, MS degree in Industrial Engineering, and BS degree in Chemical Engineering from University at Buffalo — The State University of New York, USA. His teaching and research interests are in the areas of Mathematical Programming, Manufacturing Systems Modeling and Analysis, Operations and Supply Chain Management, and Project Management. His publications have appeared in *European Journal of Operational Research, Information Systems Frontiers Journal, INFORMS Journal on Computing, International Journal of Mathematics in Operational Research, International Journal of Operational Research, International Journal of Production Economics, International Journal of Production Research, International Transactions in Operational Research, Journal of the Operational Research Society, Management Science, Multi-Criteria Decision Analysis, Operations Management Review, Operations Research,* and *WSEAS Transactions on Mathematics.* He serves/has served as a Reviewer and/or ad-hoc Editorial Team Member for many of these, as well as other journals, and for government agencies.

Mark H. Karwan is the Praxair Professor in Operations Research at the Department of Industrial and Systems Engineering at University at Buffalo — The State University of New York, USA, where he has taught for 39 years. He has broad expertise in the area of mathematical programming — modeling and algorithmic development. His 31 PhD students have been guided in areas of algorithmic development in integer programming, multiple criteria decision making and 'mixed' areas such as integer/nonlinear or integer/multi-criteria. His 100+ publications show diverse application areas such as logistics, production planning under real time pricing, capacitated lot-sizing, hazardous waste routing and security, and military path planning. Techniques to solve these problems come from the fields of linear, nonlinear and integer programming. Funding has come from NSF, ONR and industry. Prof. Karwan's industry consulting experience has largely been in the industrial gas industry and concerned with all areas of production planning, routing, forecasting and energy use planning and in supporting corporate contracts in military operations research focused on logistics and dynamic resource allocation. He has won multiple teaching awards including the (SUNY) Chancellor's Award for Excellence in Teaching. His research interests include Discrete Optimization, Multiple Criteria Decision Making, Multilevel Systems, Vehicle Routing and Scheduling, Visual Search, and Industrial Inspection.

Preface

In this book, we present a generalized framework for formulating hard combinatorial optimization problems (COPs) as polynomial-sized linear programs. Hence, the book offers a new proof of the equality of the computational complexity classes "P" and "NP". The basic model and its theoretical foundation are developed using the Traveling Salesman Problem (TSP) as an illustration. Then, our proposed generalized framework is presented and illustrated using the TSP also, as well as other well-known hard COPs. The main idea of our approach is to model COPs as flow problems over an assignment-problem (AP) graph. Our variables represent flows over doublets and triplets of arcs of the underlying graph, enabling an inductive path-theoretic argument towards proving that the proposed LP polytope has integral extrema. In the case of the TSP, the doublets and triplets of arcs respectively model doublets and triplets of travel legs, and we show that each extreme point of the resulting LP polytope corresponds to a TSP tour. Although the proposed model draws from the developments in Diaby (2006b; 2007; 2010a; 2010b), the book is fully self-contained, and does not require any familiarity with those previous developments.

There are some negative claims that we know of that have been made (through the internet, and in anonymous reviews, respectively) in direct connection to our proposed modeling approach. All of these claims have to do with relaxations of the models in Diaby (2006b; 2007) specifically. These claims are discussed briefly in the introduction chapter, and in complete detail in the appendix. Also, focusing on the TSP, we provide detailed reasons why the existing *extended formulations* "barriers" (Yannakakis (1991); Fiorini *et al.* (2011; 2012)) are not applicable to our proposed LP model. Specifically, we show in Chapters 5 and 6 that, in the case of the TSP, our proposed model: (1) is non-symmetric; (2) cannot be extended into a symmetric model using the two-indexed (city-to-city) variables that are traditionally used in defining the standard (i.e., conventional) TSP polytope; (3) does not project to the standard TSP polytope in a well-defined sense; (4) cannot be extended into (and hence, cannot lead to) a polytope which projects to the TSP polytope in a well-defined sense.

Although not reported in this book (the focus of which is on theory), our initial empirical testing on hundreds of problems has been consistent with our theoretical developments.

Acknowledgments

We wish to thank our wives (Mrs. Fatou Diaby and Mrs. Sabina Karwan) and our families who supported and encouraged us in this endeavor and throughout our professional careers in spite of all the time it took away from them. Without them our lives would have no real meaning. We are eternally grateful for their constant love.

Chapter 1

Introduction

1. Overview

In this book, we present a generalized framework for formulating hard combinatorial optimization problems (COPs) as polynomial-sized linear programs. The perspective adopted is a substantial departure from the traditional ones that have been used. It rests upon our realization that many of the well-known hard COPs (starting with the Traveling Salesman Problem (TSP)) can be modeled as Assignment Problems (APs) with side/complicating constraints (as described in Chapters 4 and 7). Our overall idea is to use a flow perspective over an AP-based graph. The novel feature in our modeling is that our variables represent "joint flows" on doublets and triplets of arcs of this graph. Hence, in essence, enough information is "built" into the variables themselves that the enforcement of the "complicating constraints" can be done implicitly (in essence; by simply setting appropriate variables to zero). A general sense of the idea may be gained by considering the standard AP/Bipartite Matching Problem (BMP). A graphical illustration of the traditional integer programming (IP)/linear programming (LP) modeling perspective for this problem is illustrated in Figure 1.1. The fact that each node in this graph is isolated reflects what we refer to as a "memorylessness" feature in the traditional IP/LP modeling of this problem. By "memorylessness" we are not referring to independence properties such as often used in probability theory or dynamic programming contexts for example. Rather, we are referring to the fact that knowledge of a particular assignment decision provides no information about the

1

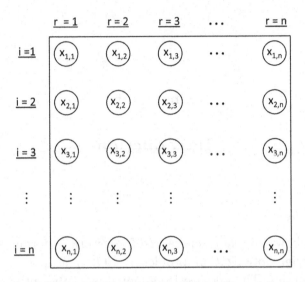

Figure 1.1. Modeling variables in the traditional IP modeling approach.

remaining decisions. In the case of the TSP, this refers to the fact that the decision made at a given stage of travel (i.e., the value of a given variable/node) in the traditional IP modeling approach) basically provides no information about the decision made at any of the other stages of travel. In other words, in the case of the TSP, the "memorylessness" we mean refers, literally, to the traveler not remembering which cities he/she has already visited while in the process of deciding which city to visit next (which could result in subtours in his/her overall travel).

The "memorylessness" described above in relation to Figure 1.1 can be remedied to some extent by using arcs to represent the decision variables, as illustrated in Figure 1.2. For example, if the problem context called for precluding decision "$x_{1,3}$" (in Figure 1.1) from being made *if* decision, say "$x_{2,2}$", has been made (and vice versa), this "complicating" requirement could be implicitly enforced (i.e., without the need for an explicit constraint) in the model based on Figure 1.2, by simply not having an arc linking nodes "$x_{2,2}$" and "$x_{1,3}$" of Figure 1.2, whereas it would require an explicit constraint in the traditional formulation approach based on Figure 1.1. We will

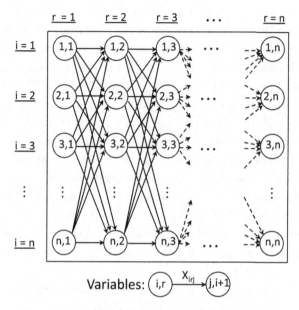

Figure 1.2. Improved choice of decision variables.

later see that in Figure 1.2, a desired solution (a tour, in the case of the TSP) would be one unit of flow "traveling" through the network in a connected manner from "stage" $r = 1$ to "stage" $r = n$ with no "level", i, repeated. Our breakthrough is in the finding that by using variables representing triplets of arcs (not necessarily connected), it is possible to formulate a model whose feasible set consists of points corresponding to convex combinations of such desired solutions only.

As far as we know, the first well-publicized attempt at formulating the TSP as a polynomial-sized LP is that of Swart (1986/1987). Not having/having had access to Swart's paper, we do not know the specifics of his model. However, the widely accepted view is that its non-validity was proved by Yannakakis (1991). The developments in Yannakakis (1991) are based on the stipulations that an LP model under consideration be symmetric and also project to the conventional TSP polytope. More recently, Fiorini *et al.* (2011; 2012) have removed the stipulation of symmetry. However, their developments still require that a model under consideration project to the conventional TSP polytope. In Chapters 5–6, we provide detailed reasons why the developments in Yannakakis (1991), and Fiorini *et al.* (2011; 2012), respectively, are not applicable to our proposed model. Specifically, we show that our basic (TSP) model: (1) is non-symmetric; (2) cannot be extended into a symmetric model using the two-indexed (city-to-city) variables that are traditionally used in defining the conventional ("standard") TSP polytope; (3) does not project to the conventional TSP polytope in a well-defined/non-ambiguous and non-degenerate/meaningful sense; (4) cannot be extended into (and hence, cannot lead to) a polytope which projects to the conventional TSP polytope in a well-defined/non-ambiguous and non-degenerate/meaningful sense. A shorter version of the material developed in Chapter 6 is also available in Diaby and Karwan (2015).

We know of three reports (by the same author) with negative claims which have been publicized having some relation to the modeling approach used in this book. There is a counter-example claim in Hofman (2006) which has to do with the relaxation of the model in Diaby (2006b) suggested in Diaby (2006a, p. 20, "Proposition 6"). There is another counter-example claim (Hofman (2008)) which pertains to a simplification of the model in Diaby (2007) discussed in Diaby (2008). Indeed, further checking revealed that each of the two relaxed models in question (i.e., Diaby (2006a), Diaby (2008)) *does* reduce to a model from which the variables representing the triplets of "travel legs" can be eliminated, and that there were flawed developments in both of the papers concerned (specifically, "Proposition 6" for Diaby (2006a), and Theorem 25 and Corollary 26 for Diaby (2008)). However, these flaws are not applicable to the published, peer-reviewed papers dealing with the "full" models (Diaby (2006b; 2007)). Hence, the counter-example claims may have had some merit, but they are applicable to the relaxations to which they pertain *only*. In the appendix to this book, we give complete details on the reasons why the model from which the variables representing triplets of "travel legs" are relaxed does not work, and why the proof logic used in this book (Chapter 3) cannot be applied to that model. The third claim of Hofman (Hofman (2007)) is of a general nature. Its premise is that an integral polytope with an exponential number of vertices cannot be completely described using a polynomially-bounded number of linear constraints (see Hofman (2007, p. 3)). However, that premise is a contradiction of many well-researched polytopes such as the Assignment Polytope (which is integral, has $n!$ extreme points (where n is the number of assignments), and is completely described by $2n$ linear constraints (see Burkard *et al.* (2007, pp. 24–26), Schrijver (1986, pp. 108–110), among others), the Transportation Polytope (see Bazaraa *et al.* (2010, pp. 513–535)), and the general Min-Cost Network Flow Polytope (see Ahuja *et al.* (1993, pp. 294–449), or Bazaraa *et al.* (2010, pp. 453–493), for example), and some non-network flow-based integral polytopes characterized in Nemhauser and Wolsey (1988, pp. 535–607), and Schrijver (1986, pp. 266–338), among others.

Clearly, this book offers a new proof of the equality of the complexity classes "*P*" and "*NP*". Note however, that our modeling and proofs do not rely on any computational complexity argument, and that hence, the "*P* = *NP*" affirmation which results from them is only incidental to the developments in the book. A distinguishing feature of the modeling approach relative to work specifically focused on the "*P* versus NP" question in general is its unifying nature. Our approach is not an isolated, *ad hoc* approach which is applicable to only a particular NP-Complete problem (such as the TSP, for example). As we show in Chapters 4 and 7 of this book, although our approach is developed using the framework of the TSP, it has natural analogs for the other problems in the NP-Complete class. Note also that our model subsumes developments focused on special-cases of specific NP-Complete problems as well, such as considered for the TSP in Garey *et al.* (1976), or in Lawler *et al.* (1985), for example.

From a practical mathematical programming perspective, the computational time complexity of problems in the *NP* class has

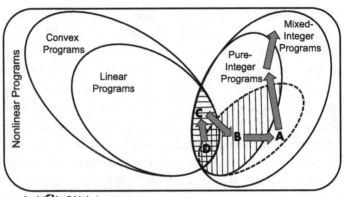

A ($\bigcirc$): 0/1 Integer programs
B (|||): Combinatorial optimization problems
C ($\equiv$): Network flow problems
D ($\boxplus$): Unit-flow network problems
(Ex: shortest path; bipartite matching)
➡: Continuum of degree of difficulty

Figure 1.3. Illustration of the continuum of degree of difficulty for ILP problems.

been viewed as being on a continuum ranging from low-degree-polynomially bounded to increasingly-higher degrees of polynomials and a shift to exponential difficulty (see Papadimitriou and Steiglitz (1982, pp. 383–398), for example). This is illustrated for integer linear programming (ILP) problems in particular in Figure 1.3. Clearly, a mixed-integer linear program (MILP) can be transformed into a pure-integer 0/1 linear program (assuming the continuous variables are rationals), and therefore, into a TSP (see Lawler *et al.* (1985, pp. 61–74)). However, the size of the resulting TSP would be several orders of magnitude greater than that of the original MILP. Hence, the developments in this book (and their incidental consequence of "$P = NP$") remove the exponential shift, but do not suggest a collapse of the "continuum of difficulty", nor any change in the sequence along that continuum. In other words, our developments do not imply (or suggest) that all of the problems in the *NP* class have become equally "easy" to solve in practice. The suggestion is that, in theory, for *NP* problems, the "continuum of difficulty" actually ranges from low-degree-polynomial time complexity to increasingly-higher-degree-polynomial time complexities.

The book is organized as follows. The basic Integer Programming (IP) model focusing on the TSP is developed in Chapter 2. We show in Chapter 3 that the Linear Programming (LP) relaxation of the basic IP model has integral extrema only, each of which corresponds to exactly one TSP tour. The generic model is developed in Chapter 4. The non-symmetry of the basic (TSP) model is discussed in Chapter 5. The non-applicability of *extended formulations* work in general is developed in Chapter 6. In Chapter 7, our proposed generalized framework is illustrated using other well-known hard COPs. Finally, conclusions are discussed in Chapter 8.

For the reader who may be interested in the "P versus NP" issue only (or, perhaps, at first), the "$P = NP$" result is reached after only the first 77 pages of the book (see Remark 3.4 of Chapter 3). Various objective functions that can be applied over our proposed LP polytope are discussed in Section 5 of Chapter 3.

Because the development of our proposed overall approach is done using the TSP as the prototypical problem, we will now provide an overview of the traditional formulations which have been proposed for the TSP. These formulations generically use the natural (city-to-city) modeling variables and focus on developing explicit constraints aimed at precluding subtours from feasible solutions. Hence, we will refer to these formulations as "natural" formulations, or "subtour elimination" formulations.

2. Overview of Traditional Formulations of the TSP

A good review of natural formulations which have been proposed for the asymmetric TSP (ATSP) is given in Öncan *et al.* (2009). We will review the three most commonly known formulations (DFJ, MTZ, and SSB) and compare them with respect to problem size.

The basic traditional model for the TSP can be stated as an assignment problem with "subtour elimination" constraints. Let n denote the number of cities; x_{ij}, a binary decision variable which is equal to 1 if there is travel from city i to city j, and 0 if not; and c_{ij}, the cost of travel from city i to city j. The basic traditional model for the TSP is as follows:

(*Problem ATSP*):

Minimize:
$$\sum_{i=1}^{n}\sum_{j=1}^{n} c_{ij}x_{ij}$$
(Minimize total cost of tour)

Subject To:
$$\sum_{j=1}^{n} x_{ij} = 1; \quad i = 1,\ldots,n$$
(Leave each city exactly once)

$$\sum_{i=1}^{n} x_{ij} = 1; \quad j = 1,\ldots,n$$
(Visit each city exactly once)

$$x_{ij} \in \{0,1\}; \quad i,j = 1,\ldots,n$$
(x_{ij} is a binary decision variable)

$$\{(i,j) : x_{ij} = 1, \quad i,j = 2,\ldots,n\}$$
does not contain subtours.

A subtour results from the assignment-only model by allowing subsets of cities to form tours, which also satisfies the conditions that every city has exactly one exit (leave) and exactly one entrance (visit). Equations or inequalities which enforce the last constraint of *Problem ATSP* above are referred to as "subtour elimination constraints (SECs)". Many SECs have been proposed. The first of these are those of Dantzig, Fulkerson and Johnson (DFJ). Their formulation is given below.

$$\sum_{i,j \in S} x_{ij} = |S| - 1; \quad S \subseteq \{2,\ldots,n\}, \ 2 \le |S| \le n - 1. \qquad (1.1)$$

Constraints (1.1) are facet-defining for *Problem ATSP*. However, there is an exponential number of them. Hence, other authors have developed polynomial-sized formulations by introducing new variables. The most recognized seminal developments among these are the Miller–Tucker–Zemlin (MTZ) SECs, and the Sarin–Sherali–Bhootra (SSB) SECs.

The MTZ formulation introduces a new set of variables, u_i ($i = 2, \ldots, n$), which define the order in which vertex i is visited. So if $u_i < u_j$ then i precedes j. The subtour elimination constraints are then:

$$\begin{cases} u_i - u_j + (n-1)x_{ij} \leq n - 2; & i, j = 2, \ldots, n, \\ 1 \leq u_i \leq n - 1; & i = 2, \ldots, n. \end{cases} \tag{1.2}$$

The SSB SECs use a different set of new variables. For each node pair (i, j), they introduce a 0/1 variable, d_{ij}, which equals 1 iff vertex i precedes (not necessarily immediately) vertex j in a tour at hand. Their model incorporates the following:

$$\begin{cases} d_{ij} \geq x_{ij}; \quad i, j = 2, \ldots, n \\ \text{(if there is travel from } i \text{ to } j, \text{ then } i \text{ precedes } j\text{),} \\[1ex] d_{ij} + d_{ji} = 1; \quad i, j = 2, \ldots, n \\ \text{(either } i \text{ precedes } j \text{ or } j \text{ precedes } i\text{),} \\[1ex] d_{ij} + d_{jk} + d_{ki} \leq 2; \quad i, j, k = 2, \ldots, n \\ \text{(if } i \text{ precedes } j \text{ and } j \text{ precedes } k, \text{ then } k \text{ cannot precede } i\text{).} \end{cases}$$

$$\tag{1.3}$$

In Table 1.1, we compare the size of the three formulations and illustrate for $n = 20$.

An interesting question has been which formulation should be employed. Öncan *et al.* (2009) reviewed 24 different "subtour elimination formulations" and compared them in a relationship diagram shown in Figure 1.4. (For details on the abbreviations of methods, see Öncan *et al.* (2009).)

Table 1.1. Size comparisons for three "subtour elimination formulations".

DFJ	$O(n^2)$	$O(2^n)$	$2^n = 1{,}048{,}576$
MTZ	$O(n^2)$	$O(n^2)$	$n^2 = 400$
SSB	$O(n^2)$	$O(n^3)$	$n^3 = 8{,}000$

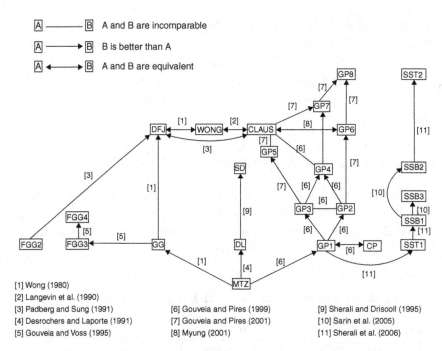

Figure 1.4. Relationship diagram comparing "subtour elimination models".

All of the "subtour elimination models" which are of polynomial size require some or all variables to be integer, and hence, are solved optimally by methods such as branch and bound. That is, the linear programming relaxation of these formulations usually results in fractional (non-TSP) solutions.

In Chapter 2 of this book, we introduce a new polynomial-sized model for the TSP which can be solved as a linear program.

3. Basic Notations, Definitions, and Assumptions for the Proposed Modeling

Assumption 1.1. *We assume, without loss of generality, that:*

(1) *The number of cities is greater than 5.*
(2) *The TSP graph is complete. (Arcs on which travel is not permitted can be handled in the optimization model by associating large ("Big-M") costs to them.)*

(3) *City "1" has been designated as the starting and ending point of the travels.*

Definition 1.1. We refer to the order in which a given city is visited after city 1 in a given TSP tour as the "time-of-travel" of that city in that TSP tour. In other words, if city i is the rth city to be visited after city 1 in a given TSP tour, then we will say that the *time-of-travel* of city i in the given tour is r.

Notation 1.1.

(1) n: Total number of cities (including the starting city);
(2) $\forall i, j \in \Omega := \{1, \ldots, n\}$, $c_{i,j}$: Cost of travel from city i to j;
(3) $m := n - 1$ (Number of cities to sequence);
(4) $M := \{2, \ldots, n\}$ (Set of cities to sequence);
(5) $S := \{1, \ldots, m\} = \{1, \ldots, n - 1\}$ (Index set for the *times-of-travel*);
(6) $R := \{1, \ldots, m - 1\} = \{1, \ldots, n - 2\}$;
(7) $\mathbb{R}$: Set of real numbers;
(8) $\mathbb{R}_+$: Set of positive real numbers;
(9) $\mathbb{R}_{\not<}$: Set of nonnegative real numbers;
(10) $\mathbb{N}$: Set of natural numbers;
(11) $\mathbb{N}_+$: Set of positive natural numbers;
(12) $\text{Conv}(\cdot)$: Convex hull of $(\cdot)$;
(13) $\text{Ext}(\cdot)$: Set of extreme points of $(\cdot)$;
(14) The notation "$\exists \langle i_1 \in A_1; \ldots; i_p \in A_p \rangle : \langle B_1; \ldots; B_q \rangle$" stands for "*There exists at least one set of p objects with one from each A_r ($r = 1, \ldots, p$), such that each expression B_s ($s = 1, \ldots, q$) is true.*" If $(i_1, \ldots, i_p)$ is unique, then we will use the notation "$\exists!$" instead of "$\exists$". Also, where that does not cause ambiguity, the brackets (one or both sets) will be omitted.
(15) All vectors are column vectors;
(16) "**0**": Vector that has every entry equal to 0;
(17) "**1**": Vector that has every entry equal to 1;
(18) $(\cdot)^T$: Transpose of vector or matrix $(\cdot)$;
(19) For two vectors a and b, $\begin{pmatrix} a \\ b \end{pmatrix} = (a^T, b^T)^T$ will be written as (a, b) whenever that is convenient and causes no confusion.

Definition 1.2 ("Standard TSP Polytope"). Let $\mathcal{A} := \{(i,j) \in \Omega^2 : i \neq j\}$ denote the set of arcs of the complete digraph on Ω. Denote the characteristic vector associated with any $F \subseteq \mathcal{A}$, by x^F (i.e., $x_{ij}^F \in \{0,1\}$ is equal to 1 iff $(i,j) \in F$). Assume (without loss of generality) that the TSP tours (defined in terms of the arcs) have been ordered, and let $\mathcal{T}_k \subset \mathcal{A}$ denote the kth tour. The "Standard TSP Polytope" (in the asymmetric case) is defined as $\mathrm{Conv}(\{x^{\mathcal{T}_k} \in \mathbb{R}^{n(n-1)}, \ k = 1, \ldots, n!\})$ (see Lawler *et al.* (1985, pp. 257–258), among others).

Chapter 2

Basic IP Model Using the TSP

1. Introduction

Our overall modeling is based on an alternate abstraction (see Borowski and Borwein (1991, p. 4)) of the TSP optimization problem in which TSP tours are modeled as Assignment (Bipartite Matching) Problem solutions. In order to be able to correctly capture TSP tour costs, the basic AP model must be reformulated in a higher-dimensional space. This reformulation is based on a path representation of the AP solutions. It may be helpful to note at this point that our modeling does not involve the *Standard TSP Polytope*.

We will start the chapter with the alternate (AP) abstraction of TSP tours. Then, we will develop the path representation which serves as the basis for the developments in the remainder of the book. Thirdly, we will discuss some further intuition of our modeling approach. Finally, our overall higher-dimensional IP reformulation of our basic AP model will be developed.

2. The "Alternate TSP Polytope": A Non-Exponential Abstraction of TSP Tours

Theorem 2.1. *Consider the TSP defined on the set of cities $\Omega :=$ $\{1, \ldots, n\}$. Assume city "1" has been designated as the starting and ending point of the travels. Let $S := \{1, \ldots, n-1\}$ denote the times-of-travel to cities "2" through "n". Then, there exists a one-to-one* *correspondence between TSP tours and extreme points of*

$$AP := \left\{ w \in \mathbb{R}_{\not<}^{(n-1)^2} : \sum_{t \in S} w_{i,t} = 1, \ \forall i \in (\Omega \backslash \{1\}); \right.$$

$$\left. \sum_{i \in (\Omega \backslash \{1\})} w_{i,t} = 1, \ \forall t \in S \right\}.$$

Proof. Using the assumption that node 1 is the starting and ending point of travel, it is trivial to construct a unique TSP tour from a given extreme point of AP, and vice versa (i.e., it is trivial to construct a unique extreme point of AP from a given TSP tour). □

Remark 2.1. *It is important to note, especially as concerns the extended formulations "barriers" works (Yannakakis (1991); Fiorini et al. (2011; 2012)), that AP is also a TSP polytope, since its extreme points correspond to TSP tours, and that its mathematical characteristics are unrelated to those of the Standard TSP Polytope.*

According to the Minkowski–Weyl Theorem (Minskowski (1910); Weyl (1935); see also Rockaffelar (1997, pp. 153–172)), every polytope can be equivalently described as the intersection of hyperplanes (ℋ-representation/external description) or as a convex combination of (a finite number of) vertices (𝒱-representation/internal description). The Standard TSP Polytope is stated in terms of its 𝒱-representation. No polynomial-sized ℋ-representation of it is known. On the other hand, AP is stated in terms of its ℋ-representation (which is well known to be of (low-degree) polynomial size (see Burkard et al. (2009)), but it is trivial to state its 𝒱-representation also.

The vertices of AP are assignment problem solutions, whereas the vertices of the Standard TSP Polytope are Hamiltonian cycles. Hence, even though the extreme points of AP and those of the Standard TSP Polytope respectively correspond to TSP tours, the two sets of extreme points are different kinds of mathematical objects, with unrelated mathematical characterizations. Hence, there does not exist any a priori mathematical relation between AP and the Standard TSP Polytope. In other words, AP and the Standard TSP Polytope are simply alternate abstractions of TSP tours. Or, put another way, AP is (simply) an alternate TSP polytope from the Standard TSP Polytope, and vice versa (i.e., that the Standard TSP Polytope is (simply) an alternate TSP polytope from AP).

Definition 2.1 ("Alternate TSP Polytope"). We refer to AP as the "Alternate TSP Polytope."

3. *"TSP Paths"*: Path Representation of TSP Tours

We reformulate AP as a generalized flow problem over the digraph illustrated in Figure 2.1. We refer to this graph as the "TSP Flow Graph (TSPFG)". The nodes of the TSPFG correspond to (city, *time-of-travel*) pairs. The arcs of the graph link nodes at consecutive *times-of-travel*. These are formalized in Notation 2.1.

Notation 2.1 (TSPFG formalisms).

(1) $\forall r \in S$, $N_r := M = \{2, \dots, n\}$ (Set of cities that *can be* visited at *time* r);

(2) $\overline{N} := \{(i, r) : r \in S, i \in N_r\}$ (Set of nodes of the TSPFG);

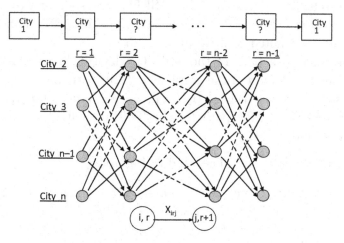

Figure 2.1. Illustration of the TSP flow graph.

(3) $\forall r \in S, \forall i \in N_r,$

$$F_r(i) := \begin{cases} \varnothing & \text{if } r = m, \\ N_{r+1} \setminus \{i\} & \text{otherwise} \end{cases}$$

(Forward star of node (i, r) of the TSPFG);

(4) $\forall r \in S, \forall i \in N_r,$

$$B_r(i) := \begin{cases} \varnothing & \text{if } r = 1, \\ N_{r-1} \setminus \{i\} & \text{otherwise} \end{cases}$$

(Backward star of node (i, r) of the TSPFG);

(5) $A : \{[i, r, j], r \in R, i \in N_r, j \in F_r(i)\}$ (Set of arcs of the TSPFG).

Remark 2.2. *It follows directly from the definitions in Notations 2.1.3–2.1.5 that:* $\forall (i, j) \in M^2, \forall r \in R, [i, r, j] \in A \Longrightarrow i \neq j.$ *That is, there is no arc in the TSPFG that connects a city with itself between consecutive times-of-travel.*

Definition 2.2 ("stages", "levels", and "TSP paths").

(1) We refer to the set of nodes of the TSPFG corresponding to a city as a "level" of the graph. (The set of *levels* is denoted M in Notation 1.1.4.)

(2) We refer to the nodes of the TSPFG corresponding to a given *time-of-travel* of the TSP as a "stage" of the graph. (The set of *stages* is denoted S in Notation 1.1.5.)

(3) We refer to a path of the TSPFG which includes exactly one node of each *level* of the graph as a "TSP path". In other words, a set of arcs, $\{[i_1, 1, i_2], [i_2, 2, i_3], \ldots, [i_{m-1}, m-1, i_m]\}$, of the TSPFG is a "TSP path (of the TSPFG)" iff $i_1, i_2, \ldots, i_m$ are pairwise-distinct.

Two *TSP paths* are illustrated in Figure 2.2.

Theorem 2.2. *There exists a one-to-one correspondence between TSP paths of the TSPFG and extreme points of the Alternate TSP Polytope (and therefore, between TSP paths of the TSPFG and TSP tours).*

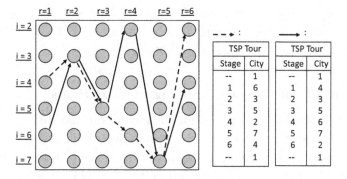

Figure 2.2. Illustration of *"TSP paths"*.

Proof.

(1) We will show that each *TSP path* corresponds to a unique extreme point of the *Alternate TSP Polytope*.

Since the number of *levels* of the TSPFG is equal to the number of *stages* of the graph, each *level* must be "visited" exactly once (at exactly one *stage*) in any given *TSP path*. For a given *TSP path*, let level i be *visited* at stage j_i. Then, by the multipartite nature of the TSPFG, we must have:

$$\forall (i, k) \in M^2 : i \neq k, j_i \neq j_k.$$

To a *TSP path* at hand, say $\mathcal{P}$, associate $\overline{w} \in \mathbb{R}^{(n-1)^2}$ such that:

$$\forall i \in M, \ \forall t \in S, \quad \overline{w}_{i,t} := \begin{cases} 0 & \text{if } t \neq j_i, \\ 1 & \text{if } t = j_i. \end{cases}$$

Then, clearly, $\overline{w}$ is an extreme point of AP, and is unique for the association to $\mathcal{P}$.

(2) We will show that each extreme point of the *Alternate TSP Polytope* corresponds to a unique *TSP path*.

Let $\overline{w} \in \text{Ext}(AP)$. For $k \in M$ and $r \in S$, let i_r be such that:

$$\overline{w}_{k,r} = \begin{cases} 0 & \text{if } k \neq i_r, \\ 1 & \text{if } k = i_r. \end{cases}$$

Since $\overline{w}$ must be such that $i_r \neq i_s$ for all $(r, s) \in S^2 : r \neq s$, one can associate the (unique) *TSP path* of the TSPFG, $\{[i_1, 1, i_2],$ $[i_2, 2, i_3], \ldots, [i_{m-1}, m - 1, i_m]\}$, to $\overline{w}$.   □

4. Intuition of the LP Modeling of *TSP Paths*

A non-mathematical intuition for the idea of our modeling approach (to be formally described in the remainder of this book) may be gained through the following observation. *Assume* that no part of whatever flow traverses a given arc of the TSPFG can subsequently "re-visit" either of the *levels* of the graph that are involved in that arc. *Then*, flow can propagate forward through the graph, *only* in a "pattern" that consists of a convex combination of *TSP paths*.

For example, consider the flow that propagates from the arc of Figure 2.3 representing travel from city 4 to city 3 at Stage 1 of the graph (i.e., "arc [4, 1, 3]"). The part of that flow that traverses arc [3, 2, 2] cannot "re-visit" (*if* the "no re-visit" condition applies) either of the levels 4, 3, or 2; the part that traverses both [3, 2, 2] and [2, 3, 6] cannot "re-visit" either of the levels 4, 3, 2, or 6; and so on. Similarly, the part of the flow from arc [4, 1, 3] that traverses arc [3, 2, 5] cannot "re-visit" either of the levels 4, 3, or 5; the part that traverses both arcs [3, 2, 5] and [5, 3, 6] cannot "re-visit" either of levels 4, 3, 5, or 6; and so on. Clearly, the resulting pattern of flow propagation would correspond to a convex combination of *TSP*

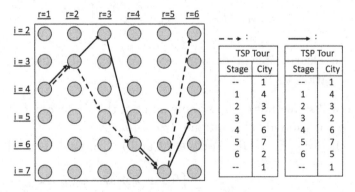

Figure 2.3. Illustration of the intuition on flow propagation patterns.

paths (if the total flow originating from arc $[4, 1, 3]$ is equal to 1), as illustrated in Figure 2.3.

In order to achieve the "no re-visit" condition described above, the "mechanism" we devise consists essentially of performing multiple, overlapping labelings ("stampings") of the flows using triplets of arcs. This results in a set of variables in our model that are defined in terms of triplets of arcs. For example, consider the total flow between arcs $[4, 1, 3]$ and $[7, 5, 2]$ of Figure 2.3. That flow is "stamped", first with arcs $[4, 1, 3]$ and $[7, 5, 2]$, and then, one additional time with whatever arcs it traverses at the intermediary stages 2, 3, and 4, respectively. Clearly, the "no re-visit" condition for arc pair $[4, 1, 3]$ and $[7, 5, 2]$ can be enforced by simply not allowing any intermediary "stamping" that involves either of the levels 4, 3, 7, or 2 of the TSPFG. Our modeling approach consists in essence of performing this "stamping" scheme for all pairs of arcs of the TSPFG, while stipulating that the total flow between any two arcs must also "visit" every *level* of the TSPFG that is not involved in the two arcs.

A key conclusion which can be drawn from the above discussion is that, *provided* the "no re-visits" requirements are properly enforced, every positive variable (as defined by arc triplets) in a solution to our LP model (to be developed in the remainder of this book) must lie on one or more *TSP paths.* In addition, proving that a solution to the model consists only of a convex combination of *TSP paths* will allow us to conclude that every extreme point of the model represents a *TSP path.*

Looking well ahead, our developments to follow will prove that feasible solutions to our proposed LP model will have the general structure illustrated in Example 3.2 of Chapter 3.

5. Integer Programming (IP) Model of *TSP Paths*

The constraint set of our proposed overall linear programming (LP) model (discussed in Chapter 3) is the same as that discussed in this section, except that the integrality requirements on the variables are relaxed (in the LP). Our reason for focusing on the IP model discussed here first is several fold, but in particular: (1) Showing that the IP model does not exclude any TSP tour means that the LP model does not exclude any TSP tour; (2) The special structure to the solutions of the LP model that makes the LP model "work" is most easily grasped by focusing on integral solutions of the model first.

We will first state the IP model and illustrate each of its *classes* of variables and constraints, respectively. Then, we will discuss the special-structure that is induced for feasible solutions.

5.1. *Modeling variables*

Notation 2.2 ("Complex" Flow Variables).

(1) $\forall (p, r, s) \in R^3 : r < p < s,\ \forall (i, j, u, v, k, t) \in (N_r, F_r(i), N_p, F_p(u), N_s, F_s(k))$, we denote the amount of flow in the TSPFG that propagates from arc $[i, r, j]$ onto arc $[k, s, t]$, via arc $[u, p, v]$, as $z_{[i,r,j][u,p,v][k,s,t]}$.

(2) $\forall (r, s) \in R^2 : r < s,\ \forall (i, j, k, t) \in (N_r, F_r(i), N_s, F_s(k))$, we denote the amount of flow in the TSPFG that propagates from arc $[i, r, j]$ onto arc $[k, s, t]$, as $y_{[i,r,j][k,s,t]}$.

Hence, our modeling variables are respectively comprised of doublets and triplets of arcs of the TSPFG, as illustrated in Figure 2.4.

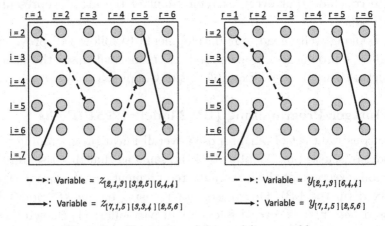

Figure 2.4. Illustration of the modeling variables.

5.2. *Model constraints*

Our IP model of *TSP paths* consists of the six classes of constraints (excluding the integrality requirements) described below. We will subsequently fully explain each of these.

"Initial flow" constraint

$$\sum_{i \in N_1} \sum_{j \in F_1(i)} \sum_{t \in F_2(j)} \sum_{v \in F_3(t)} z_{[i,1,j][j,2,t][t,3,v]} = 1. \tag{2.1}$$

"Generalized Kirchhoff Equations (GKE)"

$$\sum_{v \in B_p(u)} z_{[i,r,j][k,s,t][v,p-1,u]} - \sum_{v \in F_p(u)} z_{[i,r,j][k,s,t][u,p,v]} = 0;$$

$$\left([i,r,j],\ [k,s,t] \in A;\ (u,p) \in \overline{N}\right) : r < s < p - 1, \tag{2.2}$$

$$\sum_{v \in B_p(u)} z_{[i,r,j][v,p-1,u][k,s,t]} - \sum_{v \in F_p(u)} z_{[i,r,j][u,p,v][k,s,t]} = 0;$$

$$\left([i,r,j],\ [k,s,t] \in A;\ (u,p) \in \overline{N}\right) : r + 1 < p < s, \tag{2.3}$$

$$\sum_{v \in B_p(u)} z_{[v,p-1,u][i,r,j][k,s,t]} - \sum_{v \in F_p(u)} z_{[u,p,v][i,r,j][k,s,t]} = 0;$$

$$\left([i,r,j],\ [k,s,t] \in A;\ (u,p) \in \overline{N}\right) : 1 < p < r < s. \tag{2.4}$$

"Consistency" constraints

$$y_{[i,r,j][k,s,t]} - \sum_{v \in F_{s+1}(t)} z_{[i,r,j][k,s,t][t,s+1,v]} = 0;$$

$$[i,r,j],\ [k,s,t] \in A :\ r < s < m - 1, \tag{2.5}$$

$$y_{[i,r,j][k,s,t]} - \sum_{v \in F_{r+1}(j)} z_{[i,r,j][j,r+1,v][k,s,t]} = 0;$$

$$[i,r,j],\ [k,s,t] \in A :\ r < s - 1, \tag{2.6}$$

$$y_{[i,r,j][k,s,t]} - \sum_{v \in F_{s+1}(t)} z_{[v,r-1,i][i,r,j][k,s,t]} = 0;$$

$$[i,r,j], \ [k,s,t] \in A : \ 1 < r < s. \tag{2.7}$$

"Visit Requirements" constraints

$$y_{[i,r,j][k,s,t]} - \sum_{p \in R : p < r} \sum_{v \in F_p(u)} z_{[u,p,v][i,r,j][k,s,t]}$$

$$- \sum_{p \in R : r < p < s} \sum_{v \in B_p(u)} z_{[i,r,j][v,p,u][k,s,t]}$$

$$- \sum_{p \in R : p > s} \sum_{v \in B_p(u)} z_{[i,r,j][k,s,t][v,p,u]} = 0;$$

$$[i,r,j], \ [k,s,t] \in A : \ r < s, \ \ u \in M \backslash \{i,j,k,t\}. \tag{2.8}$$

"No flow-break" constraints

$$z_{[i,r,j][k,s,t][u,p,v]} = 0; \quad [i,r,j],[k,s,t],[u,p,v] \in A : \ (r < s < p;$$
$$((s = r+1 \text{ and } k \neq j) \ \text{ or } \ (p = s+1 \text{ and } u \neq t))). \tag{2.9}$$

"No re-visit" constraints

$$z_{[i,r,j][k,s,t][u,p,v]} = 0; \quad [i,r,j],[k,s,t],[u,p,v] \in A :$$
$$((i = k) \text{ or } (i = t) \text{ or } (i = u) \text{ or } (i = v) \text{ or}$$
$$((s \neq r+1) \text{ and } (j = k)) \text{ or } (j = t) \text{ or } (j = u) \text{ or } (j = v) \text{ or}$$
$$(k = u) \text{ or } (k = v) \text{ or}$$
$$((p \neq s+1) \text{ and } (t = u)) \text{ or } (t = v)). \tag{2.10}$$

Integrality requirements

$$z_{[i,r,j][k,s,t][u,p,v]} \in \{0,1\};$$
$$[i,r,j], \ [k,s,t], \ [u,p,v] \in A : r < s < p, \tag{2.11}$$

$$y_{[i,r,j][k,s,t]} \in \{0,1\}; \quad [i,r,j], [k,s,t] \in A : r < s. \tag{2.12}$$

Remark 2.3. *The following observations may be useful in relation to model* (2.1)–(2.12):

(1) *From Notation 2.1, we have that* $\forall(i,j) \in M^2, \forall r \in S,\ (j \in F_r(i) \Longleftrightarrow j \neq i$ *(provided* $F_r(i) \neq \varnothing$*)) and* $(j \in B_r(i) \Longleftrightarrow j \neq i$ *(provided* $B_r(i) \neq \varnothing$*)). The index sets of summations in the constraints* (2.1)–(2.8) *can be simplified accordingly. Note however, that the flow graphs for COPs other than the TSP (see Chapters 4 and 7) would be different in general from the one specified in Notation 2.1. Expressing the constraints* (2.1)–(2.8) *in terms of the formal graph notations makes the statements of these constraints generic in the sense that they do not need to be changed when the applicable flow graph is changed (see Chapters 4 and 7).*
(2) *The integrality requirements* (2.11)–(2.12) *reduce to simple nonnegativity constraints in the LP relaxation of the model.*

The *initial flow* constraint (2.1) is illustrated in Figures 2.5 and 2.6. It initiates one unit of flow at *Stage 1* of the TSPFG.

Constraints (2.2)–(2.4) generalize the Kirchhoff Equations for "regular" network flow problems (see Bazaraa *et al.*, 2010, p. 454), and are illustrated in Figure 2.7. They ensure the connectedness and balance of the "forward", "intermediate", and "backward" flow propagations, respectively, for given pairs of arcs of the TSPFG.

$$\sum_{i \in N_1} \sum_{j \in F_1(i)} \sum_{t \in F_2(j)} \sum_{v \in F_3(t)} z_{(i,1,j)(j,2,t)(t,3,v)} = 1$$

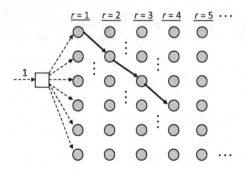

$\longrightarrow$: Example of a **term** of the summation

Figure 2.5. Variables in the *initial flow* constraint.

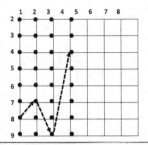

○ Define $S = \{2, 3, 4, 5, 6, 7, 8, 9\}$;

○ Writing arcs using triplets of indices, the constraint is:
$$\sum_{i,j,k,t \in S: i \neq j \neq k \neq t} \left(z_{[i,1,j][j,2,k][k,3,t]} \right) = 1$$

○ "Singling out" $z_{[8,1,7][7,2,9][9,3,4]}$ for example, the constraint can be written as:

$$z_{[8,1,7][7,2,9][9,3,4]} + \sum_{i,j,k,t \in S: (i \neq j \neq k \neq t;\ (i,j,k,t) \neq (8,7,9,4))} \left(z_{[i,1,j][j,2,k][k,3,t]} \right) = 1$$

Figure 2.6. Illustration of the *initial flow* constraint.

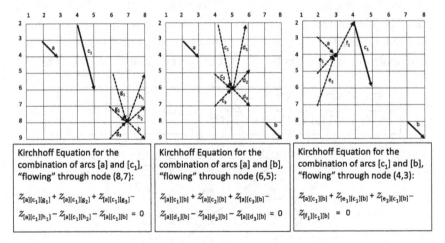

Kirchhoff Equation for the combination of arcs [a] and [c_1], "flowing" through node (8,7):	Kirchhoff Equation for the combination of arcs [a] and [b], "flowing" through node (6,5):	Kirchhoff Equation for the combination of arcs [c_1] and [b], "flowing" through node (4,3):
$z_{[a][c_1][g_1]} + z_{[a][c_1][g_2]} + z_{[a][c_1][g_3]} -$ $z_{[a][c_1][h_1]} - z_{[a][c_1][h_2]} - z_{[a][c_1][b]} = 0$	$z_{[a][c_1][b]} + z_{[a][c_2][b]} + z_{[a][c_3][b]} -$ $z_{[a][d_1][b]} - z_{[a][d_2][b]} - z_{[a][d_3][b]} = 0$	$z_{[a][c_1][b]} + z_{[e_1][c_1][b]} + z_{[e_2][c_1][b]} -$ $z_{[f_1][c_1][b]} = 0$

Figure 2.7. Illustration of the GKEs.

Constraints (2.2) stipulate that the total amount of flow from *arc* $[i, r, j]$ that propagates *forward* through *arc* $[k, s, t]$ and into (downstream) *node* $(u, p+1)$ is equal to the amount of flow from *arc* $[i, r, j]$ that propagates through *arc* $[k, s, t]$ and subsequently leaves *node* $(u, p+1)$. Constraints (2.3) stipulate that the total amount of flow from *arc* $[i, r, j]$ that enters (intermediate) *node* $(u, p+1)$ before propagating on to *arc* $[k, s, t]$ is equal to the amount of flow from *arc* $[i, r, j]$ that leaves *node* $(u, p+1)$ before propagating on to *arc* $[k, s, t]$. Constraints (2.4) stipulate that the total amount of flow that enters (upstream) node (u, p) and subsequently propagates onto arc $[k, s, t]$ via arc $[i, r, j]$ is equal to the amount of flow that leaves node (u, p) and subsequently propagates onto arc $[k, s, t]$ via arc $[i, r, j]$.

Alternatively, one can describe the GKE's using the concept of a "focus node". What gives rise to the GKE is the requirement that the "joint flow" for any pair of arcs must be balanced through any node that is chosen as the *focus node*. For example, consider the left-hand-side picture of Figure 2.7: The *focus node* is node $(8, 7)$; The *joint flow* considered pertains to arcs "a" (i.e., arc $[3, 2, 4]$) and "c_1" (i.e., arc $[2, 4, 6]$). The corresponding GKE says that sum of the z-variables respectively comprising arcs "a", "c_1", and one of the arcs entering *focus node* $(8, 7)$ must be equal to the sum of the z-variables respectively comprising arcs "a", "c_1", and one of the arcs leaving *focus node* $(8, 7)$. The middle picture of Figure 2.7 illustrates this condition for the case where the *focus node* lies (at a stage) between (the stages of) the arcs whose *joint flow* is being considered. The right-hand-side picture of the figure illustrates the GKE constraint for the case where the *focus node* lies to the right of the arcs whose *joint flow* is being considered.

Note that GKE's based on a given pair of arcs $[i, r, j]$ and $[k, s, t]$ cannot be written for the nodes upon which those arcs are incident (i.e., nodes (i, r), $(j, r+1)$, (k, s), and $(t, s+1)$). The "mass conservations" implicit in the GKE's are enforced for those nodes (i.e., (i, r), $(j, r+1)$, (k, s), and $(t, s+1)$) by the *consistency* constraints (2.5)–(2.7), as illustrated in Figure 2.8. Constraints (2.5) and (2.7) together ensure the consistency between the *backward* and *forward* flow propagations for given pairs of arcs of the TSPFG. The combination of

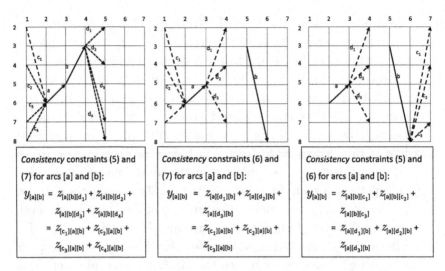

Figure 2.8. Illustration of the *consistency* constraints.

constraints (2.5) and (2.6) ensure the consistency between the *intermediate* and *forward* flow propagations. Constraints (2.6) and (2.7) enforce the consistency between the *intermediate* and *backward* flow propagations.

The connectedness implicit in the GKE's is ensured for nodes (i, r), $(j, r+1)$, (k, s), and $(t, s+1)$, respectively, by the *no-flow-break* constraints (2.9) illustrated in Figure 2.9 (along with the constraints (constraints 2.10) that enforce the *no-revisit* requirement of the TSP for the cities comprising a given modeling variable).

Constraints (2.8) ensure that the flow between any pair of arcs is part of a flow pattern that "visits" every level of the TSPFG. This is illustrated in Figure 2.10 for arcs "a" (i.e., [3,3,4]) and "b" (i.e., [8,7,9]), and *level*/city 6. The requirement that needs to be enforced for these is that the (whole) *joint flow* for arcs "a" and "b" must flow to *level*/city 6. This is modeled by constraints (2.8) for the two arcs ("a" and "b") and the city (city 6) under consideration. The figure illustrates the stipulation of constraints (2.8) that the sum of the z-variables respectively comprising arcs "a", "b", and one of the other arcs shown in the figure be equal to the total *joint flow* for arcs "a", "b". Note that in formulating this stipulation, if city 6 is "visited" before the first city in the sequence of the cities

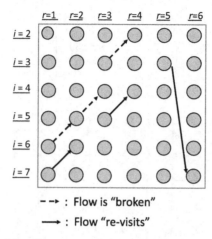

Figure 2.9. Illustration of the implicitly-zero variables.

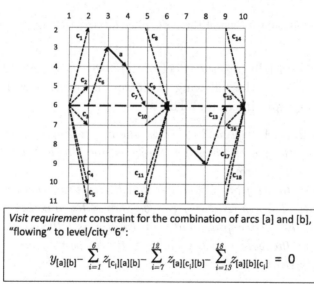

Figure 2.10. Illustration of the *visit requirements* constraints.

comprising arcs "a" and "b", then it is sufficient to consider only arcs "leaving" city 6, and that, otherwise, it is sufficient to consider only arcs "entering" city 6. It is important to note also that in this scheme, the "flow" connectivities are ensured by the GKE's.

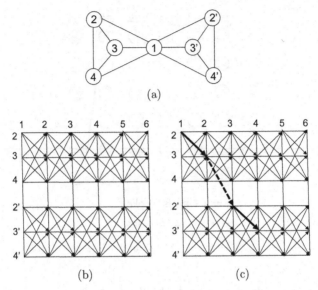

Figure 2.11. First illustration of the *visit requirements* constraints for an infeasible TSP.

A further insight into the effect of the *visit constraints* (2.8) may be gained through the following two examples dealing with special (incomplete-graph, infeasible) TSP's.

Example 2.1. *A "special" TSP is shown in Figure 2.11.a. The TSPFG (omitting arcs on which travel is not allowed) is shown in Graph (b) of the figure.*

It is easy to see from Graph (a) that there is no feasible solution to this TSP. This infeasibility can be shown for our LP model by focusing on the visit requirements constraints as follows.

Consider the variables associated with the pair of arcs ([2, 1, 3], [2′, 3, 3′]) illustrated in Graph (c). We have:

- *Consistency constraints (2.6)* $\Longrightarrow$

$$y_{[2,1,3][2′,3,3′]} = z_{[2,1,3][3,2,2′][2′,3,3′]}. \tag{2.13}$$

- *Because arc [3, 2, 2′] does not exist in the TSPFG (Graph (b) of the figure):*

$$z_{[2,1,3][3,2,2′][2′,3,3′]} = 0. \tag{2.14}$$

- (2.13) *and* (2.14) $\Longrightarrow$

$$y_{[2,1,3][2',3,3']} = z_{[2,1,3][3,2,2'][2',3,3']} = 0. \tag{2.15}$$

- *It is easy to see that a similar reasoning as in* (2.13)–(2.15) *can be applied for every pair of arcs* $([i,r,j],\ [k,r+1,t])$ *with* $(i,j) \in \{2,3,4\}^2$ *and* $(k,t) \in \{2',3,4'\}^2$. *Hence, because of the "flow" connectedness of our model (due to the GKE's* (2.2)–(2.4) *and the no-flow-break constraints* (2.9)), *there can be no "flow" between any arc* $([i,r,j]$ *with* $(i,j) \in \{2,3,4\}^2$ *and any other arc* $[k,s,t])$ *with* $(k,t) \in \{2',3,4'\}^2$ *in a feasible solution to our model for the TSP of Graph* (a).
- *Hence, the visit requirements constraints* (2.8) *cannot be satisfied for the special TSP of Figure* 2.11.

Example 2.2. *Consider the "special" TSP shown in Figure* 2.12.a. *The TSPFG for this TSP is shown in Graph* (b) *of the figure.*

The infeasibility of our model for this special TSP can also be seen from the impossibility of satisfying the visit requirements constraints

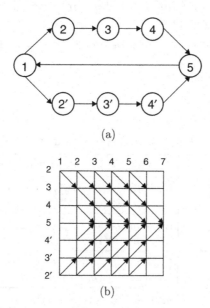

(a)

(b)

Figure 2.12. Second illustration of the *visit requirements* constraints for an infeasible TSP.

for any two of the arcs involving levels 2, 3, *or* 4, *and any one of the levels* 2′, 3′, *or* 4′, *and vice versa (i.e., for any two of the arcs involving levels* 2′, 3′, *or* 4′, *and any one of the levels* 2, 3, *or* 4), *in particular.*

Remark 2.4. *The nodes* (i,r), $(j, r+1)$, (k,s), *and* $(t, s+1)$ *may be thought of as "boundary nodes", with respect to the GKE's based on the pair of arcs* $[i,r,j]$ *and* $[k,s,t]$. *Hence, the "consistency" and "no-flow-break" constraints can be thought of as enforcing "boundary conditions" of the GKE's.*

The following theorem shows that the size of our model is polynomial in the number (n) of cities, although the model is very-large-scale. Recall that we generally use $m = n - 1$ in our analysis, since we can assume we start and end at city 1 and model our problem using the remaining $n - 1$ cities/stages.

Theorem 2.3.

(1) *The number of (non-implicitly-zero) variables in the system* (2.1)–(2.8) *is* $O(m^9)$,
(2) *The number of constraints in the system* (2.1)–(2.8) *is* $O(m^8)$.

Proof. (i) *Statement* (1).

We have:

Table 2.1. Number of z-variables.

$z_{[i,r,j][j,r+1,t][t,r+2,v]}$	$(m-3)\,(_mP_4) = m(m-1)(m-2)(m-3)^2$
$z_{[i,r,j][j,r+1,t][u,p,v]}$: $p > r+2$	$(m-3)(m-4)\,(_mP_5)\,/2 = m(m-1)(m-2)$ $\times\,(m-3)^2(m-4)^2/2$
$z_{[i,r,j][k,p-1,u][u,p,v]}$: $p > r+2$	$(m-3)(m-4)\,(_mP_5)\,/2 = m(m-1)(m-2)$ $\times\,(m-3)^2(m-4)^2/2$
$z_{[i,r,j][k,s,u][u,p,v]}$: $s > r+1; p > s+1$	$(_{m-3}P_3)\,(_mP_6)\,/6 = m(m-1)(m-2)(m-3)^2$ $\times\,(m-4)^2(m-5)^2/6$
Total, $\xi_z =$	$m(m-1)(m-2)(m-3)^2(m^4 - 18m^3 + 127m^2$ $- 408m + 502)/6$

Table 2.2. Number of y-variables.

$y_{[i,r,j][j,r+1,t]}$	$(m-2)\,(_mP_3) = m(m-1)(m-2)^2$
$y_{[i,r,j][k,s,t]} : s > r+1$	$(m-2)(m-3)\,(_mP_4)\,/2 = m(m-1)(m-2)^2(m-3)^2/2$
Total, $\xi_y =$	$m(m-1)(m-2)^2(m^2-6m+11)/2$

Hence, the numbers of y- and z-variables in the system are:

$$\begin{cases} \xi_y := m(m-1)(m-2)^2(m^2-6m+11)/2, \\ \xi_z := m(m-1)(m-2)(m-3)^2 \\ \qquad \times\,(m^4-18m^3+127m^2-408m+502)/6. \end{cases}$$

(ii) *Statement* (2).

We have:

Table 2.3. Number of constraints.

Consistency constraints:

$[i,1,j],[j,2,t]$	$_mP_3 =$	$m(m-1)(m-2)$
$[i,m-2,j],[j,m-1,t]$	$_mP_3 =$	$m(m-1)(m-2)$
$[i,r,j],[j,r+1,t]$: $2 \le r \le m-3$	$2(m-4)(_mP_3) =$	$2m(m-1)(m-2)(m-4)$
$[i,1,j],[k,s,t]$: $3 \le s \le m-2$	$2(_mP_5) =$	$2m(m-1)(m-2)(m-3)$ $\times\,(m-4)$
$[i,r,j],[k,m-1,t]$: $2 \le r \le m-3$	$2(_mP_5) =$	$2m(m-1)(m-2)(m-3)$ $\times\,(m-4)$
$[i,1,j],[k,m-1,t]$	$_mP_4 =$	$m(m-1)(m-2)(m-3)$
$[i,r,j],[k,s,t]$: $2 \le r \le m-4$; $r+2 \le s \le m-2$	$3(_mP_6)/2 =$	$3m(m-1)(m-2)(m-3)$ $\times\,(m-4)(m-5)/2$

Visit Requirements constraints:

$[i,r,j],[j,r+1,t]$: $1 \le r \le m-2$	$(m-2)\,(_mP_4) =$	$m(m-1)(m-2)^2(m-3)$
$[i,r,j],[k,s,t]$: $1 \le r \le m-3$; $r+2 \le s \le m-1$	$(m-2)(m-3)$ $(_mP_5)\,/2 =$	$m(m-1)(m-2)^2(m-3)^2$ $\times\,(m-4)/2$

(*Continued*)

Table 2.3. (*Continued*)

Kirchhoff Equations:

$[i, 1, j], [j, 2, t]$	$_mP_5 =$	$m(m-1)(m-2)(m-3)(m-4)$
$[i, m-2, j], [j, m-1, t]$	$_mP_5 =$	$m(m-1)(m-2)(m-3)(m-4)$
$[i, 2, j], [j, 3, t]$	$(m-5)(_mP_4) =$	$m(m-1)(m-2)(m-3)(m-5)$
$[i, m-3, j], [j, m-2, t]$	$(m-5)(_mP_4) =$	$m(m-1)(m-2)(m-3)(m-5)$
$[i, r, j], [j, r+1, t]$: $3 \le r \le m-4$	$(m-5)(m-6)$ $\times (_mP_4) =$	$m(m-1)(m-2)(m-3)$ $\times (m-5)(m-6)$
$[i, 1, j], [k, s, t]$: $3 \le s \le m-2$	$(m-4)(_mP_6) =$	$m(m-1)(m-2)(m-3)$ $\times (m-4)^2(m-5)$
$[i, r, j], [k, m-1, t]$: $2 \le r \le m-3$	$(m-4)(_mP_6) =$	$m(m-1)(m-2)(m-3)$ $\times (m-4)^2(m-5)$
$[i, 1, j], [k, m-1, t]$	$(m-4)(_mP_5) =$	$m(m-1)(m-2)(m-3)$ $\times (m-4)^2$
$[i, 2, j], [k, s, t]$: $4 \le s \le m-3$	$(m-6)^2(_mP_5) =$	$m(m-1)(m-2)(m-3)$ $\times (m-4)(m-6)^2$
$[i, r, j], [k, m-2, t]$: $3 \le r \le m-4$	$(m-6)^2(_mP_5) =$	$m(m-1)(m-2)(m-3)$ $\times (m-4)(m-6)^2$
$[i, 2, j], [k, m-2, t]$	$(m-6)(_mP_5) =$	$m(m-1)(m-2)(m-3)$ $\times (m-4)(m-6)$
$[i, r, j], [k, s, t]$: $3 \le r \le m-5$; $r+2 \le s \le m-3$	$(m-6)^2(m-7)$ $\times (_mP_5)/2 =$	$m(m-1)(m-2)(m-3)$ $\times (m-4)(m-6)^2(m-7)/2$

Hence, the number of constraints in the system is:

$$\eta := 1 + m\,(m-1)\,(m-2)\,(m-3)$$
$$\times (m^4 - 14m^3 + 80m^2 - 213m + 222)/2.$$ $\square$

Notation 2.3. *In the remainder of this book, we will denote the numbers of y-variables, z-variables, and constraints of Model (2.1)–(2.12) by ξ_y, ξ_z, and η, respectively, as given in Tables 2.1, 2.2, and 2.3, respectively.*

Hence, our model is polynomial-sized, although it is very-large-scale as can be seen from Table 2.4, where the last column is the

Table 2.4. Illustration of the large-scale nature of the model.

n	y-variables	z-variables	Total	$(n-1)^9$	Ratio
6	540	480	1,020	1.953125E+06	5.222400000E−04
7	2,640	6,120	8,760	1.007770E+07	8.692463039E−04
8	9,450	53,760	63,210	4.035361E+07	1.566402726E−03
9	27,216	344,400	371,616	1.342177E+08	2.768754959E−03
10	67,032	1,681,344	1,748,376	3.874205E+08	4.512864058E−03
11	146,880	6,597,360	6,744,240	1.000000E+09	6.744240000E−03
12	294,030	21,795,840	22,089,870	2.357948E+09	9.368261257E−03
13	547,800	62,833,320	63,381,120	5.159780E+09	1.228368568E−02
14	962,676	162,333,600	163,296,276	1.060450E+10	1.539877275E−02
15	1,611,792	383,447,064	3.850589E+08	2.066105E+10	1.863694807E−02
20	12,750,102	1.12760E+10	1.12887E+10	3.226877E+11	3.498350685E−02
30	200,670,372	8.98515E+11	8.98716E+11	1.450715E+13	6.194987095E−02
60	1.74443E+10	9.01126E+14	9.01144E+14	8.662996E+15	1.040221563E−01
100	4.20738E+11	1.15351E+17	1.15352E+17	9.135172E+17	1.262719266E−01

(*Continued*)

Table 2.4. (*Continued*)

n	y-variables	z-variables	Total	$(n-1)^9$	Ratio
150	5.07924E+12	5.02234E+18	5.02235E+18	3.619732E+19	1.387491859E−01
300	3.44321E+14	2.907E+21	2.907E+21	1.910032E+22	1.521963611E−01
400	1.96247E+15	3.99107E+22	3.99107E+22	2.563044E+23	1.557159059E−01
1,000	4.91559E+17	1.60761E+26	1.60761E+26	9.910359E+26	1.622149172E−01
10,000	4.99151E+23	1.66068E+35	1.66068E+35	991004E+35	1.662171514E−01
100,000	4.99915E+29	1.66607E+44	1.66607E+44	9.999100E+44	1.666216715E−01
1E+06	4.99992E+35	1.66661E+53	1.66661E+53	9.999910E+53	1.666621667E−01
1E+07	5.0000E+41	1.66666E+62	1.66666E+62	9.999991E+62	1.666662167E−01
1E+08	5.0000E+47	1.66667E+71	1.66667E+71	9.999999E+71	1.666666217E−01
1E+09	5.0000E+53	1.66667E+80	1.66667E+80	1.000000E+81	1.666666622E−01

ratio of the total number of variables (i.e., the fourth column) to the number m^9 (i.e., the fifth column). As can be seen from the table the ratio converges to the fraction $\frac{1}{6} = 0.1\overline{66}$, which is consistent with the expression for ξ_z above (see Table 2.1).

Remark 2.5. *Throughout the remainder of this book, we will write the solution vector to our proposed model as "(y, z)" whenever that is convenient and causes no ambiguity, as indicated in Chapter 1 (see Notation 1.1.19).*

5.3. *"Derived" consistency constraints*

In this section, we will discuss a more expanded form of the *consistency* constraints (2.5)–(2.7) that is derived from the combination of constraints (2.5)–(2.7) with the GKE constraints (2.2)–(2.4). We will use these "derived consistency constraints" in place of constraints (2.5)–(2.7) in the remainder of this book, due to the fact that they simplify the proofs. Note however, that in our empirical testing (which is not reported in this book), we have used constraints (2.5)–(2.7) only. For a given pair of arcs $[i, r, j]$ and $[k, s, t]$ of the TSPFG, the *derived* constraints are simply explicit statements of the consistency requirements for each of the stages of the graph. For example, consider consistency constraints (2.5) and (2.7). These constraints ensure that the total flow based on the given pair of arcs $[i, r, j]$ and $[k, s, t]$ that propagates to stage $(s + 1)$ (accounted by constraints (2.5)) is *consistent* with the total flow based on these arcs that propagates from stage $(r - 1)$ (accounted by constraints (2.7)). Note that the GKE constraints (2.2) for $[i, r, j]$ and $[k, s, t]$ in turn, ensure the consistency between the total flows at stage $(s + 1)$ and any stage $t : t > (s + 1)$, and that the GKE constraints (2.4) for $[i, r, j]$ and $[k, s, t]$ ensure the consistency between the total flows at stage $(r - 1)$ and any stage $p : p < (r - 1)$. Hence, it is valid to add constraints that say that the total flow based on a given pair of arcs, $[i, r, j]$ and $[k, s, t]$, must be consistent between any two stages p and t such that $p \leq (r - 1)$ and $t \geq (s + 1)$. A similar reasoning can be applied to cases involving constraints (2.6), to arrive at the general statement that the total flow based on a given pair of arcs, $[i, r, j]$ and $[k, s, t]$, must be consistent between any two stages p and t of the TSPFG. This is formalized in the following theorem.

Theorem 2.4. *The following constraints are valid for* $\{(y,z) \in \mathbb{R}^{(\xi_y + \xi_z)} : (y,z) \ satisfies \ (2.1)–(2.12)\}$:

$$y_{[i,r,j][k,s,t]} - \sum_{u \in N_p} \sum_{v \in F_p(u)} z_{[i,r,j][k,s,t][u,p,v]} = 0;$$

$$([i,r,j], \ [k,s,t] \in A, \ p \in R) : r < s < p, \qquad (2.16)$$

$$y_{[i,r,j][k,s,t]} - \sum_{u \in N_p} \sum_{v \in F_p(u)} z_{[i,r,j][u,p,v][k,s,t]} = 0;$$

$$([i,r,j], \ [k,s,t] \in A, \ p \in R) : r < p < s, \qquad (2.17)$$

$$y_{[i,r,j][k,s,t]} - \sum_{u \in N_p} \sum_{v \in F_p(u)} z_{[u,p,v][i,r,j][k,s,t]} = 0;$$

$$([i,r,j], \ [k,s,t] \in A, \ p \in R) : p < r < s. \qquad (2.18)$$

Proof. The theorem follows directly from the combination of the GKE constraints (2.2)–(2.4) and the *consistency* constraints (2.5)–(2.7). We will illustrate the proof with the details for the "expanded" *consistency* constraints (2.16).

GKE constraints (2.2) for $([i,r,j], \ [k,s,t] \in A; \ (u, s+2) \in \overline{N}) :$ $r < s < m - 2$ gives:

$$\sum_{v \in F_{s+2}(u)} z_{[i,r,j][k,s,t][u,s+2,v]}$$

$$= \sum_{v \in B_{s+2}(u)} z_{[i,r,j][k,s,t][v,s+1,u]}$$

$$= z_{[i,r,j][k,s,t][t,s+1,u]} \ (\text{using the } \textit{no-flow-break} \text{ constraints (2.9)}).$$
$$(2.19)$$

Summing both sides of (2.19) over the u's gives:

$$\sum_{u \in N_{s+2}} \sum_{v \in F_{s+2}(u)} z_{[i,r,j][k,s,t][u,s+2,v]}$$

$$= \sum_{u \in N_{s+2}} z_{[i,r,j][k,s,t][t,s+1,u]}$$

$$= y_{[i,r,j][k,s,t]} \text{ (using "original" } consistency \text{ constraints (2.5))}.$$

$$(2.20)$$

Summing the GKE constraints (2.2) over the u's gives:

$$\sum_{u \in N_p} \sum_{v \in F_p(u)} z_{[i,r,j][k,s,t][u,p,v]} = \sum_{u \in N_p} \sum_{v \in B_p(u)} z_{[i,r,j][k,s,t][v,p,u]};$$

$$([i,r,j],\ [k,s,t] \in A;\ p \in R) : r < s < p - 1. \tag{2.21}$$

Constraints (2.16) follow directly from the combination of (2.20) and (2.21). The proof steps for constraint sets (2.17) and (2.18) respectively, are similar. $\qquad\square$

Remark 2.6.

(1) *Constraint set (2.16)–(2.18) subsumes constraint set (2.5)–(2.7);*
(2) *Throughout the remainder of this book, by "consistency constraints" we will be referring to the derived constraints (2.16)–(2.18). That is, we will use (2.16)–(2.18) in the place of (2.5)–(2.7) in the remainder of this book.*

Definition 2.3.

(1) Let $Q_I := \{(y,z) \in \mathbb{R}^{(\xi_y + \xi_z)} : (y,z)$ satisfies (2.1)–(2.4), (2.16)–(2.18), (2.8)–(2.12)\}. We refer to $Conv(Q_I)$ as the "IP Polytope";
(2) We refer to the linear programming relaxation of Q_I as the "LP Polytope", and denote it by Q_L; i.e., $Q_L := \{(y,z) \in \mathbb{R}^{(\xi_y + \xi_z)} : (y,z)$ satisfies (2.1)–(2.4), (2.16)–(2.18), (2.8)–(2.10), and $\mathbf{0} \le (y,z) \le \mathbf{1}\}$;
(3) We will alternatively refer to a point of the *LP Polytope* as a "solution to the LP", or a "LP solution", or a "LP solution instance" in the remainder of the book, whenever convenient.

6. Structure of the IP Polytope

The theorem below shows the unique mathematical characterization of *TSP paths* in our modeling approach. For the IP model, the characterization is that each triplet of arcs along a given *TSP path* of the TSPFG corresponds to a z-variable that is equal to 1, and that similarly, every pair of arcs along the path corresponds to a y-variable that is equal to 1 in the model.

Theorem 2.5. $(y, z) \in Q_I \iff \exists! m\text{-tuple } (i_r \in N_r, \ r = 1, \ldots, m)$:

(1) $\forall (p, q) \in (S, S\backslash\{p\}), \ i_p \neq i_q$; (*In words, no level/city is repeated in the m-tuple.*)

(2) $\forall t \in M, \ \exists p \in S : i_p = t$; (*In words, every level/city is included in the m-tuple.*)

(3)

$$y_{[a,r,b][c,s,d]} = \begin{cases} 1 & \text{for } (r, s) \in R^2 : r < s, \text{ and} \\ & (a, b, c, d) = (i_r, i_{r+1}, i_s, i_{s+1}); \\ 0 & \text{otherwise.} \end{cases}$$

(4)

$$z_{[a,r,b][c,s,d][e,p,f]} = \begin{cases} 1 & \text{for } (p, r, s) \in R^3 : r < s < p, \text{ and} \\ & (a, b, c, d, e, f) = (i_r, i_{r+1}, i_s, i_{s+1}, i_p, i_{p+1}); \\ 0 & \text{otherwise.} \end{cases}$$

Proof.

(a) $\implies$:

Let $(y, z) \in Q_I$. The proof has several steps, as follows.

Step 0. We will first derive additional constraints that are *valid* for Q_I (and also, for Q_L).

(**0.1**) We will show that:

$$\sum_{u_r \in N_r} \sum_{v_r \in F_r(u_r)} \sum_{u_s \in N_s} \sum_{v_s \in F_s(u_s)} \sum_{u_t \in N_t} \sum_{v_t \in F_t(u_t)} z_{[u_r,r,v_r][u_s,s,v_s][u_t,t,v_t]} = 1$$

$$\forall (r, s, t) \in R^3 : r < s < t.$$

(In words, we will show that the total of the *joint flows* involving the arcs at any given set of three *stages* is equal to 1.)

(0.1.1) *Case* 1: $(r, s, t) = (1, 2, 3)$.

We have:

$$\sum_{u_1 \in N_1} \sum_{v_1 \in F_1(u_1)} \sum_{u_2 \in N_2} \sum_{v_2 \in F_2(u_2)} \sum_{u_3 \in N_3} \sum_{v_3 \in F_3(u_3)} z_{[u_1,1,v_1][u_2,2,v_2][u_3,3,v_3]}$$

$$= \sum_{u_1 \in N_1} \sum_{v_1 \in F_1(u_1)} \sum_{v_2 \in F_2(v_1)} \sum_{v_3 \in F_3(v_2)} z_{[u_1,1,v_1][v_1,2,v_2][v_2,3,v_3]}$$

(Using *no-flow-break* constraints (2.9) (on p. 22))

$$= 1 \quad \text{(Using \textit{initial flow} constraint (2.1) (on p. 21)).} \qquad (2.22)$$

(0.1.2) *Case* 2: $(r, s) = (1, 2)$, $t > 3$.

We have:

$\forall s \in R : s > 3,$

$$\sum_{u_1 \in N_1} \sum_{v_1 \in F_1(u_1)} \sum_{u_2 \in N_2} \sum_{v_2 \in F_2(u_2)} \sum_{u_s \in N_s} \sum_{v_s \in F_s(u_s)} z_{[u_1,1,v_1][u_2,2,v_2][u_s,s,v_s]}$$

$$= \sum_{u_1 \in N_1} \sum_{v_1 \in F_1(u_1)} \sum_{u_2 \in N_2} \sum_{v_2 \in F_2(u_2)} y_{[u_1,1,v_1][u_2,2,v_2]}$$

(Using *consistency* constraints (2.16))

$$= \sum_{u_1 \in N_1} \sum_{v_1 \in F_1(u_1)} \sum_{u_2 \in N_2} \sum_{v_2 \in F_2(u_2)} \sum_{u_3 \in N_3} \sum_{v_3 \in F_3(u_3)} z_{[u_1,1,v_1][u_2,2,v_2][u_3,3,v_3]}$$

(Using (2.16))

$$= \sum_{u_1 \in N_1} \sum_{v_1 \in F_1(u_1)} \sum_{v_2 \in F_2(v_1)} \sum_{v_3 \in F_3(v_2)} z_{[u_1,1,v_1][v_1,2,v_2][v_2,3,v_3]}$$

(Using (2.9))

$$= 1 \quad \text{(Using (2.22)).} \qquad (2.23)$$

(0.1.3) *Case* 3: $r = 1, 2 < s < t$.

We have:

$\forall (r, s) \in R^2 : 2 < r < s,$

$$\sum_{u_1 \in N_1} \sum_{v_1 \in F_1(u_1)} \sum_{u_r \in N_r} \sum_{v_r \in F_r(u_r)} \sum_{u_s \in N_s} \sum_{v_s \in F_s(u_s)} z_{[u_1,1,v_1][u_r,r,v_r][u_s,s,v_s]}$$

$$= \sum_{u_1 \in N_1} \sum_{v_1 \in F_1(u_1)} \sum_{u_s \in N_s} \sum_{v_s \in F_s(u_s)} \sum_{u_r \in N_r} \sum_{v_r \in F_r(u_r)} z_{[u_1,1,v_1][u_r,r,v_r][u_s,s,v_s]}$$

(Re-arranging)

$$= \sum_{u_1 \in N_1} \sum_{v_1 \in F_1(u_1)} \sum_{u_s \in N_s} \sum_{v_s \in F_s(u_s)} y_{[u_1,1,v_1][u_s,s,v_s]}$$

(Using *consistency* constraints (2.17))

$$= \sum_{u_1 \in N_1} \sum_{v_1 \in F_1(u_1)} \sum_{u_s \in N_s} \sum_{v_s \in F_s(u_s)} \sum_{u_2 \in N_2} \sum_{v_2 \in F_2(u_2)} z_{[u_1,1,v_1][u_2,2,v_2][u_s,s,v_s]}$$

(Using (2.17))

$$= \sum_{u_1 \in N_1} \sum_{v_1 \in F_1(u_1)} \sum_{u_2 \in N_2} \sum_{v_2 \in F_2(u_2)} \sum_{u_s \in N_s} \sum_{v_s \in F_s(u_s)} z_{[u_1,1,v_1][u_2,2,v_2][u_s,s,v_s]}$$

(Rearranging)

$= 1$ (Using (2.23)).   (2.24)

(0.1.4) *Case* 4: $1 < r < s < t$.

We have:

$\forall (r, s, t) \in R^3 : 1 < r < s < t,$

$$\sum_{u_r \in N_r} \sum_{v_r \in F_r(u_r)} \sum_{u_s \in N_s} \sum_{v_s \in F_s(u_s)} \sum_{u_t \in N_t} \sum_{v_t \in F_t(u_t)} z_{[u_r,r,v_r][u_s,s,v_s][u_t,t,v_t]}$$

$$= \sum_{u_r \in N_r} \sum_{v_r \in F_r(u_r)} \sum_{u_s \in N_s} \sum_{v_s \in F_s(u_s)} y_{[u_r,r,v_r][u_s,s,v_s]}$$

(Using *consistency* constraints (2.16))

$$= \sum_{u_r \in N_r} \sum_{v_r \in F_r(u_r)} \sum_{u_s \in N_s} \sum_{v_s \in F_s(u_s)} \sum_{u_1 \in N_1} \sum_{v_1 \in F_1(u_1)} z_{[u_1,1,v_1][u_r,r,v_r][u_s,s,v_s]}$$

(Using *consistency* constraints (2.18))

$$= \sum_{u_1 \in N_1} \sum_{v_1 \in F_1(u_1)} \sum_{u_r \in N_r} \sum_{v_r \in F_r(u_r)} \sum_{u_s \in N_s} \sum_{v_s \in F_s(u_s)} z_{[u_1,1,v_1][u_r,r,v_r][u_s,s,v_s]}$$

(Re-arranging)

$$= 1 \quad \text{(Using (2.24))}. \tag{2.25}$$

(0.1.5) *Conclusion/Synthesis.*

We have: (2.22)–$(2.25) \Longrightarrow$

$\forall (r,s,t) \in R^3 : r < s < t,$

$$\sum_{u_r \in N_r} \sum_{v_r \in F_r(u_r)} \sum_{u_s \in N_s} \sum_{v_s \in F_s(u_s)} \sum_{u_t \in N_t} \sum_{v_t \in F_t(u_t)} z_{[u_r,r,v_r][u_s,s,v_s][u_t,t,v_t]} = 1.$$
$$\tag{2.26}$$

(0.2) We will show that:

$$\sum_{u_r \in N_r} \sum_{v_r \in F_r(u_r)} \sum_{u_s \in N_s} \sum_{v_s \in F_s(u_s)} y_{[u_r,r,v_r][u_s,s,v_s]} = 1 \ \forall (r,s) \in R^2 : r < s.$$

(In words, we will show that the total of the *joint flows* involving the arcs at any given set of two *stages* is equal to 1.)

(0.2.1) *Case* 1: $(r,s) = (1, m-1)$.

We have:

$$\sum_{u_1 \in N_1} \sum_{v_1 \in F_1(u_1)} \sum_{u_{m-1} \in N_{m-1}} \sum_{v_{m-1} \in F_{m-1}(u_{m-1})} y_{[u_1,1,v_1][u_{m-1},m-1,v_{m-1}]}$$

$$= \sum_{u_1 \in N_1} \sum_{v_1 \in F_1(u_1)} \sum_{u_{m-1} \in N_{m-1}} \sum_{v_{m-1} \in F_{m-1}(u_{m-1})}$$

$$\sum_{u_r \in N_r} \sum_{v_r \in F_r(u_r)} z_{[u_1,1,v_1][u_r,r,v_r][u_{m-1},m-1,v_{m-1}]},$$

$$r \in R \backslash \{1, m-1\} \quad \text{(Using (2.17))}$$

$$= \sum_{u_1 \in N_1} \sum_{v_1 \in F_1(u_1)} \sum_{u_r \in N_r} \sum_{v_r \in F_r(u_r)}$$

$$\sum_{u_{m-1} \in N_{m-1}} \sum_{v_{m-1} \in F_{m-1}(u_{m-1})} z_{[u_1,1,v_1][u_r,r,v_r][u_{m-1},m-1,v_{m-1}]},$$

$$r \in R \backslash \{1, m-1\} \quad \text{(Rearranging)}$$

$$= 1 \quad \text{(Using (2.26))}. \tag{2.27}$$

(0.2.2) *Case* 2: $1 < r < s$.

Then: (2.18) and (2.26) $\implies$

$$\forall (r,s) \in R^2 : 1 < r < s,$$

$$\sum_{u_r \in N_r} \sum_{v_r \in F_r(u_r)} \sum_{u_s \in N_s} \sum_{v_s \in F_s(u_s)} y_{[u_r,r,v_r][u_s,s,v_s]}$$

$$= \sum_{u_r \in N_r} \sum_{v_r \in F_r(u_r)} \sum_{u_s \in N_s} \sum_{v_s \in F_s(u_s)} \sum_{u_1 \in N_1} \sum_{v_1 \in F_1(u_1)} z_{[u_1,1,v_1][u_r,r,v_r][u_s,s,v_s]}$$

(Using (2.18))

$$= \sum_{u_1 \in N_1} \sum_{v_1 \in F_1(u_1)} \sum_{u_r \in N_r} \sum_{v_r \in F_r(u_r)} \sum_{u_s \in N_s} \sum_{v_s \in F_s(u_s)} z_{[u_1,1,v_1][u_r,r,v_r][u_s,s,v_s]}$$

(Rearranging)

$$= 1 \quad \text{(Using (2.26))}. \tag{2.28}$$

(0.2.3) *Case* 3: $r < s < m - 1$.

Then: (2.16) and (2.26) $\implies$

$$\forall (r,s) \in R^2 : r < s < m-1,$$

$$\sum_{u_r \in N_r} \sum_{v_r \in F_r(u_r)} \sum_{u_s \in N_s} \sum_{v_s \in F_s(u_s)} y_{[u_r,r,v_r][u_s,s,v_s]}$$

$\forall (r,s) \in R^2 : r < s < m - 1,$

$$\sum_{u_r \in N_r} \sum_{v_r \in F_r(u_r)} \sum_{u_s \in N_s} \sum_{v_s \in F_s(u_s)} y_{[u_r,r,v_r][u_s,s,v_s]}$$

$$= \sum_{u_r \in N_r} \sum_{v_r \in F_r(u_r)} \sum_{u_s \in N_s} \sum_{v_s \in F_s(u_s)}$$

$$\sum_{u_{m-1} \in N_{m-1}} \sum_{v_{m-1} \in F_{m-1}(u_{m-1})} z_{[u_r,r,v_r][u_s,s,v_s][u_{m-1},m-1,v_{m-1}]}$$

(Using (2.16))

$$= 1 \quad \text{(Using (2.26))}. \tag{2.29}$$

(0.2.4) *Conclusion/Synthesis*:

We have: (2.27)–(2.29) $\implies$

$$\forall (r,s) \in R^2 : r < s, \quad \sum_{u_r \in N_r} \sum_{v_r \in F_r(u_r)} \sum_{u_s \in N_s} \sum_{v_s \in F_s(u_s)} y_{[u_r,r,v_r][u_s,s,v_s]} = 1. \tag{2.30}$$

Step 1: We will show that there exists exactly one m-tuple ($i_r \in M$, $r = 1, \ldots, m$) such that:

$$z_{[i_1,1,i_2][i_2,2,i_3][i_p,p,i_{p+1}]} = 1, \quad p = 3, \ldots, m - 1. \tag{2.31}$$

The proof is as follows.

(1.1) First, *integrality* constraints (2.11) and *initial flow* constraint (2.1) $\implies$

$\exists! 4$-tuple ($i_r \in M, r = 1, \ldots, 4$) such that:

$$1 = \sum_{u_1 \in N_1} \sum_{u_2 \in F_1(i)} \sum_{u_3 \in F_2(j)} \sum_{u_4 \in F_3(t)} z_{[u_1,1,u_2][u_2,2,u_3][u_3,3,u_4]}$$

$$= z_{[i_1,1,i_2][i_2,2,i_3][i_3,3,i_4]}. \tag{2.32}$$

(1.2) Second, using (2.32) and (2.11), and applying GKE constraints (2.2) for arc pair $([i_1, 1, i_2], [i_2, 2, i_3])$ at *stages* 3 through $m - 1$ successively, we get:

$$\exists!(m - 4)\text{-tuple } (i_r \in M, r = 5, \ldots, m) : z_{[i_1,1,i_2][i_2,2,i_3][i_p,p,i_{p+1}]} = 1,$$

for $p = 4, \ldots, m - 1$.   (2.33)

(1.3) Finally, (2.31) follows directly from the combination of (2.32) and (2.33).

Step 2: *Statement* (1) *of Theorem* 2.5.

We will show that the m-tuple $(i_r \in M, r = 1, \ldots, m)$ of *Step* 1 is such that:

$$\forall (p, q) \in S : p \neq q, \; i_p \neq i_q.$$ (2.34)

The proof is as follows.

(2.1) First, we have:

(i) $i_1 \neq i_2 \neq i_3 \neq i_4$ (Using (2.32) and *no re-visit* constraints (2.10));

(ii) $\forall p \in S : p > 3, \; i_p \notin \{i_1, i_2, i_3\}$ (Using (2.33) and (2.10));

(iii) $y_{[i_1,1,i_2][i_2,2,i_3]} = \displaystyle\sum_{u \in N_3} \sum_{v \in F_3(u)} z_{[i_1,1,i_2][i_2,2,i_3][u,3,v]}$

(Using (2.16))

$= z_{[i_1,1,i_2][i_2,2,i_3][i_3,3,i_4]}$

(Using (2.11) and (2.32))

$= 1$ (Using (2.32)).

                                                                                 (2.35)

(2.2) Now, let $(r, s) \in R^2 : 2 < r < s$.

Assume $i_r = i_s = j$.

Then, we must have:

$$(2.35(i)) \text{ and } (2.35(ii)) \implies \begin{cases} \text{(i)} \quad r > 3; \text{ and} \\ \\ \text{(ii)} \quad s > 3. \end{cases} \tag{2.36}$$

$$(2.32), (2.33), \text{ and } (2.36) \implies \begin{cases} \text{(i)} \quad z_{[i_1,1,i_2][i_2,2,i_3][i_{r-1},r-1,j]} = 1; \text{ and} \\ \\ \text{(ii)} \quad z_{[i_1,1,i_2][i_2,2,i_3][i_{s-1},s-1,j]} = 1. \end{cases} \tag{2.37}$$

Also, using (2.36) and (2.37), *visit requirements* constraint ((2.8); p. 22) for arc pair $([i_1,1,i_2], [i_2,2,i_3])$ and *level j* gives:

$$
\begin{aligned}
0 = {}& y_{[i_1,1,i_2][i_2,2,i_3]} - \sum_{p \in S: p > 2} \sum_{u \in (N_p \setminus \{i_1,i_2,i_3,j\})} z_{[i_1,1,i_2][i_2,2,i_3][u,p,j]} \\
= {}& y_{[i_1,1,i_2][i_2,2,i_3]} - z_{[i_1,1,i_2][i_2,2,i_3][i_{r-1},r-1,j]} - z_{[i_1,1,i_2][i_2,2,i_3][i_{s-1},s-1,j]} \\
& - \sum_{(p,u) \, \in \, ((S\setminus\{1,2\}, \, N_p\setminus\{i_1,i_2,i_3,j\})\setminus\{(r-1,i_{r-1}),(s-1,i_{s-1})\})} z_{[i_1,1,i_2][i_2,2,i_3][u,p,j]} \\
= {}& 1 - 1 - 1 - \sum_{(p,u) \, \in \, ((S\setminus\{1,2\}, \, N_p\setminus\{i_1,i_2,i_3,j\})\setminus\{(r-1,i_{r-1}),(s-1,i_{s-1})\})} z_{[i_1,1,i_2][i_2,2,i_3][u,p,j]} \\
= {}& -1 - \sum_{(p,u) \, \in \, ((S\setminus\{1,2\}, \, N_p\setminus\{i_1,i_2,i_3,j\})\setminus\{(r-1,i_{r-1}),(s-1,i_{s-1})\})} z_{[i_1,1,i_2][i_2,2,i_3][u,p,j]}.
\end{aligned}
\tag{2.38}
$$

Note that (2.11) $\implies$

$$\sum_{(p,u) \, \in \, ((S\setminus\{1,2\}, \, N_p\setminus\{i_1,i_2,i_3,j\}) \setminus \{(r-1,i_{r-1}),(s-1,i_{s-1})\})} z_{[i_1,1,i_2][i_2,2,i_3][u,p,j]} \geq 0. \tag{2.39}$$

Hence, (2.38) cannot be satisfied.

Hence, we must have:

$$\forall (p,q) \in (S\setminus\{i_1,i_2,i_3\})^2 : p \neq q, \ i_p \neq i_q. \tag{2.40}$$

(2.3) *Synthesis.*

(2.34) follows directly from the combination of (2.40), (2.35(ii)), and (2.35(i))

Step 3: *Statement (2) of Theorem 2.5.*

It follows directly from (2.34) and the fact that $|M| = |S| = m$ that the m-tuple $(i_r \in M, r = 1, \ldots, m)$ of *Step 1* satisfies statement (2) of Theorem 2.5.

Step 4: *Statement (3) of Theorem 2.5.*

We will show that the m-tuple $(i_r \in M, r = 1, \ldots, m)$ of *Step 1* is such that:

$$\forall (r, s) \in R^2 : r < s, \; y_{[i_r,r,i_{r+1}][i_s,s,i_{s+1}]} = 1. \tag{2.41}$$

The proof is as follows.

(4.1) (2.31) and *integrality* constraints (2.12) $\Longrightarrow$

$$\begin{cases} \text{(i)} \;\; y_{[i_1,1,i_2][i_2,2,i_3]} = 1 \;\; \text{(Using (2.16))}; \\[2mm] \text{(ii)} \;\; \forall r \in R : r > 2, \; y_{[i_1,1,i_2][i_r,r,i_{r+1}]} = 1 \;\; \text{(Using (2.17))}; \\[2mm] \text{(iii)} \;\; \forall r \in R : r > 2, \; y_{[i_2,2,i_3][i_r,r,i_{r+1}]} = 1 \;\; \text{(Using (2.16))}. \end{cases} \tag{2.42}$$

(2.42(i)), (2.42(ii)), (2.17), and *integrality* constraints (2.11) $\Longrightarrow$

$$\forall (r, s) \in R^2 : 1 < r < s, \; \exists!(u_r, u_{r+1}) \in M^2 :$$
$$z_{[i_1,1,i_2][u_r,r,u_{r+1}][i_s,s,i_{s+1}]} = 1. \tag{2.43}$$

(4.2) Let $(r, s) \in R^2 : 1 < r < s$, and $(j_r, j_{r+1}) \in M^2$ be such that

$$z_{[i_1,1,i_2][j_r,r,j_{r+1}][i_s,s,i_{s+1}]} = 1. \tag{2.44}$$

(4.2.1) First, using (2.12), (2.16) and (2.44) $\Longrightarrow$

$$y_{[i_1,1,i_2][j_r,r,j_{r+1}]} = 1. \tag{2.45}$$

(**4.2.2**) Second, assume $(j_r, j_{r+1}) \neq (i_r, i_{r+1})$.

Then, using (2.45) and (2.42ii), we get:

$$\sum_{u_1 \in N_1} \sum_{u_2 \in F_1(u_1)} \sum_{u_r \in N_r} \sum_{u_{r+1} \in F_r(u_r)} y_{[u_1,1,u_2][u_r,r,u_{r+1}]}$$

$$\geq y_{[i_1,1,i_2][j_r,r,j_{r+1}]} + y_{[i_1,1,i_2][i_r,r,i_{r+1}]}$$

$$+ \sum_{\substack{(u_1,u_2)\in(N_1,\ F_1(u_1)): \\ (u_1,u_2)\neq(i_1,i_2)}} \sum_{\substack{(u_r,u_{r+1})\in \\ ((N_r,\ F_r(u_r))\setminus\{(j_r,j_{r+1}),\ (i_r,i_{r+1})\})}}$$

$$\times y_{[u_1,1,u_2][u_r,r,u_{r+1}]}$$

$$\geq 2. \tag{2.46}$$

(2.46) contradicts (2.30).

(**4.2.3**) Hence, we must have:

$$(j_r, j_{r+1}) = (i_r, i_{r+1}). \tag{2.47}$$

Also:

$$(2.47) \implies z_{[i_1,1,i_2][i_r,r,i_{r+1}][i_s,s,i_{s+1}]} = 1, \text{ and} \tag{2.48}$$

$$(2.48) \implies y_{[i_r,r,i_{r+1}][i_s,s,i_{s+1}]} = 1. \text{ (Using (2.18) and (2.12))} \tag{2.49}$$

(**4.3**) Finally, it follows from the combination of (2.42), (2.43) and (2.47)–(2.49), that:

$$\forall (r, s) \in R^2 : r < s, \ y_{[i_r,r,i_{r+1}][i_s,s,i_{s+1}]} = 1. \tag{2.50}$$

(**4.4**) In addition (to *Steps* 4.1–4.3), we also have:

(2.30) and (2.50) $\implies$

$\forall (r, s) \in R^2 : r < s,$

$\forall (u_r, u_{r+1}, u_s, u_{s+1}) \in M^4 : (u_r, u_{r+1}, u_s, u_{s+1}) \neq (i_r, i_{r+1}, i_s, i_{s+1}),$

$$y_{[u_r,r,u_{r+1}][u_s,s,u_{s+1}]} = 0. \tag{2.51}$$

It follows directly from the combination of (2.50) and (2.51), that the m-tuple ($i_r \in M$, $r = 1, \ldots, m$) of *Step 1* satisfies statement (iii) of Theorem 2.5.

Step 5: *Statement* (iv) *of Theorem* 2.5.

We will show that the m-tuple ($i_r \in M$, $r = 1, \ldots, m$) of *Step 1* is such that:

$$\forall (p, r, s) \in R^2 : p < r < s, \ z_{[i_p,p,i_{p+1}][i_r,r,i_{r+1}][i_s,s,i_{s+1}]} = 1. \quad (2.52)$$

The proof is as follows.

(5.1) Let $(r, s) \in R^2 : r < s$.

Then, using (2.11) and (2.16)–(2.18), we have:

$$(2.50) \Longrightarrow \begin{cases} \text{(i)} \ \ \forall p \in R : p < r, \\ \qquad \exists! (u_p, u_{p+1}) \in M^2 : z_{[u_p,p,u_{p+1}][i_r,r,i_{r+1}][i_s,s,i_{s+1}]} = 1, \\ \\ \text{(ii)} \ \ \forall p \in R : r < p < s, \\ \qquad \exists! (u_p, u_{p+1}) \in M^2 : z_{[i_r,r,i_{r+1}][u_p,p,u_{p+1}][i_s,s,i_{s+1}]} = 1, \\ \\ \text{(iii)} \ \ \forall p \in R : s < p, \\ \qquad \exists! (u_p, u_{p+1}) \in M^2 : z_{[i_r,r,i_{r+1}][i_s,s,i_{s+1}][u_p,p,u_{p+1}]} \\ \qquad = 1. \end{cases}$$

$$(2.53)$$

(5.2)

(5.2.1) *Case 1:*

Let $(p, r, s) \in R^2 : p < r < s$ and $(u_p, u_{p+1}) \in M^2$ be such that:

$$z_{[u_p,p,u_{p+1}][i_r,r,i_{r+1}][i_s,s,i_{s+1}]} = 1. \quad (2.54)$$

Then:

(2.12), (2.16), and (2.54) $\Longrightarrow$

$$y_{[u_p,p,u_{p+1}][i_r,r,i_{r+1}]} = 1. \quad (2.55)$$

Using (2.51) and (2.55), we must have:

$$(u_p, u_{p+1}) = (i_p, i_{p+1}). \quad (2.56)$$

(5.2.2) *Case 2:*

Let $(p, r, s) \in R^2 : r < p < s$, and $(u_p, u_{p+1}) \in M^2$ be such that:

$$z_{[i_r,r,i_{r+1}][u_p,u_{p+1}][i_s,s,i_{s+1}]} = 1. \tag{2.57}$$

Then:

(2.12), (2.16), and (2.57) $\Longrightarrow$

$$y_{[u_p,u_{p+1}][i_s,s,i_{s+1}]} = 1. \tag{2.58}$$

Using (2.51) and (2.58), we must have:

$$(u_p, u_{p+1}) = (i_p, i_{p+1}). \tag{2.59}$$

(5.2.3) *Case 3:*

Let $(p, r, s) \in R^2 : r < s < p$, and $(u_p, u_{p+1}) \in M^2$ be such that:

$$z_{[i_r,r,i_{r+1}][i_s,s,i_{s+1}][u_p,u_{p+1}]} = 1. \tag{2.60}$$

Then:

(2.12), (2.17), and (2.60) $\Longrightarrow$

$$y_{[i_r,r,i_{r+1}][u_p,p,u_{p+1}]} = 1. \tag{2.61}$$

Using (2.51) and (2.61), we must have:

$$(u_p, u_{p+1}) = (i_p, i_{p+1}). \tag{2.62}$$

(5.2.4) *Conclusion/Synthesis.*

(2.52) follows directly from the combination of (2.56), (2.59), and (2.62).

(5.3) In addition (to *Steps* 5.1 and 5.2), we have:

(2.52) and (2.26) $\Longrightarrow$

$$\forall (p, r, s) \in R^2 : p < r < s, \ \forall (u_p, u_{p+1}, u_r, u_{r+1}, u_s, u_{s+1}) \in M^4 :$$

$$(u_p, u_{p+1}, u_r, u_{r+1}, u_s, u_{s+1}) \neq (i_p, i_{p+1}, i_r, i_{r+1}, i_s, i_{s+1}),$$

$$z_{[i_p,p,i_{p+1}][i_r,r,i_{r+1}][i_s,s,i_{s+1}]} = 0. \tag{2.63}$$

It follows directly from the combination of (2.62) and (2.63), that the m-tuple $(i_r \in M, r = 1, \ldots, m)$ of *Step* 1 satisfies statement (iv) of Theorem 2.5.

(b) $\Longleftarrow$: Trivial.   □

The next set of results, together, establish that every integral point of our proposed linear programming polytope is an extreme point of the polytope and corresponds to exactly one TSP tour. Their importance relative to the developments to come in Chapter 3 is due to the fact that they make it sufficient to later show (in the chapter) that every point of the LP polytope is a convex combination of integral points of the polytope.

Theorem 2.6. *The following statements are true*:

(1) *There exists a one-to-one correspondence between the points of Q_I and the TSP paths (of the TSPFG)*;
(2) *There exists a one-to-one correspondence between the points of Q_I and the TSP tours.*

Proof. Statement (1) follows directly from the combination of Theorem 2.5 and Definition 2.2.3. Statement (2) follows from the combination of Statement (1) with Theorem 2.2.   □

Theorems 2.5 and 2.6 are illustrated in Figures 2.13 and 2.14.

Theorem 2.7. *Every point of Q_I is an extreme point of Q_L.*

Proof. Let $\bar{n}_z((y,z))$ and $\bar{n}_y((y,z))$ denote the numbers of positive z- and y-components of $(y,z) \in \mathbb{R}^{(\xi_y + \xi_z)} : 0 \le (y,z) \le 1$, respectively. Then, using Theorem 2.5, we have:

(a) $(y,z) \in Q_I \iff$
$$
\begin{cases}
\bar{n}_z((y,z)) = \dbinom{m-1}{3} = \dfrac{(m-1)!}{3!(m-4)!}, \\[2ex]
\bar{n}_y((y,z)) = \dbinom{m-1}{2} = \dfrac{(m-1)!}{2!(m-3)!}; \quad \text{and}
\end{cases}
$$

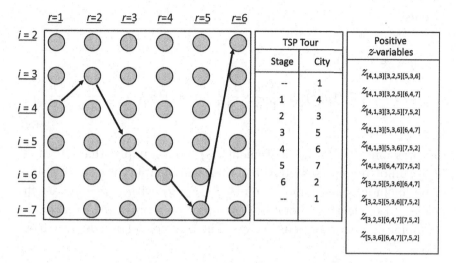

Figure 2.13. Illustration of the positive z-variables for a TSP tour.

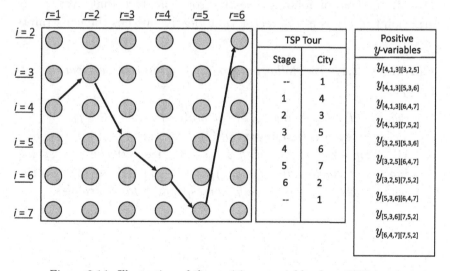

Figure 2.14. Illustration of the positive y-variables for a TSP tour.

$$(b) \quad (y,z) \in Q_L \implies \begin{cases} \overline{n}_z((y,z)) \geq \dfrac{(m-1)!}{3!(m-4)!}; \\[2ex] \overline{n}_y((y,z)) \geq \dfrac{(m-1)!}{2!(m-3)!}. \end{cases}$$

Hence,

$$(y, z) \in Q_L \backslash Q_I$$

$$\implies \left(\overline{n}_z((y, z)) > \frac{(m-1)!}{3!(m-4)!} \quad \text{or} \quad \overline{n}_y((y, z)) > \frac{(m-1)!}{2!(m-3)!} \right).$$

$$(2.64)$$

Expression (2.64) and statement (a) above imply that $(y, z) \in Q_I$ cannot be represented as a convex combination of a set of points that includes one or more points of $Q_L \backslash Q_I$ with a positive coefficient. Also, clearly, a point of Q_I cannot be represented as a convex combination of other points of Q_I. The theorem follows directly from these. $\square$

Theorem 2.8. *Every integral point of Q_L is a point of Q_I.*

Proof. The proof follows directly from the fact that every integral point of Q_L is a $0/1$ vector that clearly satisfies the constraints of Q_I. $\square$

Corollary 2.1. *The following are true of integral extreme points of Q_L:*

(1) $((\widehat{y}, \widehat{z}) \in Ext(Q_L)$ *and* $(\widehat{y}, \widehat{z})$ *integral*$) \Longleftrightarrow (\widehat{y}, \widehat{z}) \in Q_I$;

(2) *Every integral extreme point of Q_L has exactly $\overline{n}_z((y, z)) = (m-1)!/(3!(m-4)!)$ entries with values "1", respectively, corresponding to z-variables, and $\overline{n}_y((y, z)) = (m-1)!/(2!(m-3)!)$ entries with values "1", respectively, corresponding to y-variables.*

Chapter 3

Basic LP Model Using the TSP

1. Introduction

The constraint set of our proposed LP model is Q_L. In this chapter, we will first discuss some general algebraic characterizations of Q_L. Then, we will show that each extreme point of Q_L is a point of Q_I. Alternate (linear) cost functions to associate to these points so that TSP tours are correctly abstracted in the overall TSP optimization problem will also be discussed (in Section 5). The proof that all extreme points of Q_L are integral will be done in two parts. Part I (Section 3) involves characterizations of a *TSP path* of the TSPFG in terms of the variables of the LP model, and establishes that every positive variable in a feasible LP solution is contained in at least one *TSP path* (as characterized in terms of our modeling variables). In Part II (Section 4), we use the results of Part I and a *scaling* property of Q_L (developed in Section 2) to establish that a solution of the LP is "made up" of *TSP paths* (as characterized in terms of the variables) only.

Although not reported in this book, and although not serving as a "proof of validity", for each of hundreds of problems we have experimented with (an "MPS file generator" computer code is available upon request from the first author of this book), we have consistently obtained optimal (integral) solutions, consistent with the theoretical developments to follow in the remainder of this book. For a 10-city TSP, the proposed LP model has 1,748,376 variables and 1,723,681 constraints. Hence, the largest problems we were able to experiment with were 9-city TSP's (which have $371,616$ variables and $475,441$ constraints).

2. General Algebraic Characterizations of the LP Polytope

In this section, we will develop some general algebraic and geometric characterizations of Q_L.

Theorem 3.1. *The set of extreme points of Q_L is linearly independent.*

Proof. For convenience, write $x = \binom{y}{z}$ and denote the set of extreme points of Q_L as $\widehat{X} := \{\widehat{x}_1, \ldots, \widehat{x}_\nu\} = \text{Ext}(Q_L)$.

We have (see Panik (1993, pp. 235–239), among others):

$$Q_L \neq \varnothing \implies \widehat{X} \text{ is affinely independent.} \qquad (3.1)$$

Also, since $Q_L \subseteq \text{Conv}(\widehat{X} \cup \{\mathbf{0}\})$, we must have:

$$\text{Conv}(\widehat{X} \cup \{\mathbf{0}\}) \neq \varnothing. \qquad (3.2)$$

Suppose $(\widehat{x}_i, \widehat{x}_j) \in \widehat{X}^2 : \widehat{x}_i \neq \widehat{x}_j$ are such that $\widehat{x}_i = \alpha \widehat{x}_j$ (where α is a scalar). Then, clearly, $\widehat{x}_i$ and $\widehat{x}_j$ cannot both satisfy constraint (2.1) (in particular). Hence, the following is true:

$$\forall (\widehat{x}_i, \widehat{x}_j) \in \widehat{X}^2 : \widehat{x}_i \neq \widehat{x}_j, \ \{\widehat{x}_i, \widehat{x}_j\} \text{ is linearly independent.} \quad (3.3)$$

Let $\widetilde{x}$ be a conic combination of integral extreme points of Q_L with "weights" summing up to less than 1 (i.e., $\widetilde{x} = \sum_{k=1}^{p} \mu_k \widehat{x}_{i_k}$; $\mu_k > 0$ for all k; $\sum_{k=1}^{p} \mu_k < 1$; $\widehat{x}_{i_k}$ integral for all k). Then, since the integral extreme points of Q_L have 0/1 components (see Theorem 2.8), $\widetilde{x}$ cannot satisfy constraint (2.1). Hence, $\widetilde{x}$ cannot be a feasible point of Q_L. Hence, there cannot exist an extreme point of Q_L which is a conic combination of integral extreme points of Q_L with "weights" summing up to 1 or less. It follows from the combination of this with statement (3.3) and the convexity of Q_L (by which every convex combination of (any number of) integral extreme points of Q_L must be feasible for Q_L), that every member of $\widehat{X}$ is also an extreme point of $(\widehat{X} \cup \{\mathbf{0}\})$. Hence, we must have:

$$\text{Ext}(\text{Conv}(\widehat{X} \cup \{\mathbf{0}\})) = \widehat{X} \cup \{\mathbf{0}\}. \tag{3.4}$$

Now, (3.2) and (3.4) imply:

$$\widehat{X} \cup \{\mathbf{0}\} \text{ is affinely independent.} \tag{3.5}$$

From (3.5), we have that:

$$\{(\widehat{x}_1 - \mathbf{0}), \dots, (\widehat{x}_\nu - \mathbf{0})\}$$
$$= \{\widehat{x}_1, \dots, \widehat{x}_\nu\} = \widehat{X} \text{ is linearly independent.} \qquad \square$$

Definition 3.1. Let λ be a scalar on the interval $(0, 1]$. We refer to $\widetilde{Q}_L(\lambda) := \{(y, z) \in \mathbb{R}^{(\xi_y + \xi_z)} : (y, z) \text{ satisfies } (2.2)\text{--}(2.4), (2.16)\text{--}(2.18), (2.8)\text{--}(2.10), (3.6), \text{ and } \mathbf{0} \le (y, z) \le \lambda \cdot \mathbf{1}\}$ as the "λ-scaled LP polytope", where (3.6) is specified as:

$$\sum_{i \in N_1} \sum_{j \in F_1(i)} \sum_{t \in F_2(j)} \sum_{v \in F_3(t)} z_{[i,1,j][j,2,t][t,3,v]} = \lambda. \tag{3.6}$$

Remark 3.1. *The upper bounds on the y- and z-variables in Q_L and $\widetilde{Q}_L(\lambda)$, respectively, are redundant. Hence, the λ-scaled LP polytope $(0 < \lambda \le 1)$ is essentially the version of our proposed LP model in which the total flow has been scaled to λ.*

We will now show that for every $\lambda \in (0, 1]$, $\widetilde{Q}_L(\lambda)$ and Q_L have the same topological and geometric properties. We will do this by showing that for any two scalars $\lambda, \mu \in (0, 1]$, $\widetilde{Q}_L(\lambda)$ and $\widetilde{Q}_L(\mu)$ are both homeomorphic and homothetic to one another. (In other words, we will show that any two given *scalings* of the LP polytope have points that have the same "patterns"/properties (see Gamelin and Greene (1999, pp. 67–96)), and that moreover, the associated polytopes have the same "shape" (see Coxeter (1989, pp. 67–76))).

Theorem 3.2. *The following statements are true:*

(i) $\forall \lambda, \mu \in (0,1]$, $\widetilde{Q}_L(\lambda)$ and $\widetilde{Q}_L(\mu)$ are homeomorphic.
(ii) $\forall \lambda, \mu \in (0,1]$, $\widetilde{Q}_L(\lambda)$ and $\widetilde{Q}_L(\mu)$ are homothetic.

Proof. (i) It is easy to observe that constraints (2.2)–(2.4), (2.16)–(2.18), and (2.8)–(2.10) together, essentially induce "balanced" flows over the TSPFG, and that the "balance" of any given such (induced) flow is preserved under multiplication by a positive scalar. Hence,

$$\forall \alpha \in (0,1], (y,z) \in Q_L \Longleftrightarrow \alpha \cdot (y,z) \in \widetilde{Q}_L(\alpha).$$

(Alternatively, one could simply multiply each of the constraints of the LP model by α to obtain the result.)

Hence, for $\alpha \in (0,1]$, the point-to-point mapping

$$h_\alpha : Q_L \longrightarrow \widetilde{Q}_L(\alpha) \text{ with } h_\alpha((y,z)) = \alpha \cdot (y,z)$$

is bijective. Clearly h_α is bicontinuous (see Gamelin and Greene (1999, pp. 26–27), or Panik (1993, pp. 267–268)). Hence, h_α is a homeomorphism (see Gamelin and Greene (1999, pp. 27, 67), or Panik (1993, pp. 253–257)) between Q_L and $\widetilde{Q}_L(\alpha)$. Hence, $\forall(\lambda, \mu) \in (0,1]^2 : \mu \neq \lambda$, $\widetilde{Q}_L(\lambda)$ and $\widetilde{Q}_L(\mu)$ are respectively homeomorphic to Q_L. The homeomorphism between $\widetilde{Q}_L(\lambda)$ and $\widetilde{Q}_L(\mu)$ follows directly from the combination of this and the equivalence property of homeomorphisms (see Gamelin and Greene (1999, pp. 67–96)).

(ii) The statement is trivial when $\lambda = \mu$. Hence, assume (without loss of generality) that $\lambda > \mu$. Then, it follows directly from (i) above that

$$h_{\mu/\lambda} : \widetilde{Q}_L(\lambda) \longrightarrow \widetilde{Q}_L(\mu) \text{ with } h_{\mu/\lambda}((y,z)) = (\mu/\lambda) \cdot (y,z)$$

is a *dilatation* over $\widetilde{Q}_L(\lambda)$ (see Coxeter (1989, pp. 67–76)).   □

Notation 3.1. *We denote* $\widetilde{Q}_L(\lambda)$ *augmented with the zero-vector by* $\widetilde{Q}_L^0(\lambda)$ *(i.e.,* $\widetilde{Q}_L^0(\lambda) := ((\widetilde{Q}_L(\lambda)) \cup \{\mathbf{0}\}))$.

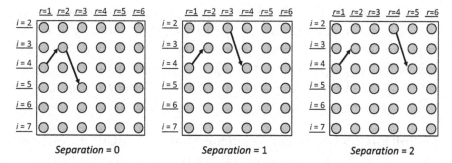

Figure 3.1. Illustration of *arc separation*.

3. "Flow" Structure of the LP Polytope

In this section, we will develop the special structure of the feasible solutions to our LP constraint set (Q_L). We will begin by offering a set of definitions and terminologies which will facilitate the remainder of the discussion.

Definition 3.2 ("Arc Separation"). For any $([i, r, j], [k, s, t]) \in A^2$: $s > r$, we refer to the quantity $(s - (r + 1)) = (s - r - 1)$ as the "arc separation (between $[i, r, j]$, and $[k, s, t]$)".

Arc Separation is illustrated in Figure 3.1.

Definition 3.3 ("(Arc) Communication").

(1) We say that two arcs of the TSPFG "2-communicate" in a given LP solution instance if and only if the y-variable corresponding to them is positive in the solution instance. In other words, arcs $[i, r, j]$ and $[k, s, t]$ of the TSPFG are said to "2-communicate" in $(y, z) \in Q_L$ iff $y_{[i,r,j][k,s,t]} > 0$.

(2) Similarly, we say that three arcs of the TSPFG "3-communicate" in a given LP solution instance if and only if the z-variable corresponding to them is positive in the solution instance. In other words, arcs $[i, r, j]$, $[k, s, t]$, and $[u, p, v]$ of the TSPFG are said to "3-communicate" in $(y, z) \in Q_L$ iff $z_{[i,r,j][k,s,t][u,p,v]} > 0$.

Definition 3.4 ("Communication path"). For $(r, s) \in R^2 : s > r+1$, we refer to the set of arcs in a path of the support graph of an LP solution instance, $([i_r, r, i_{r+1}], \ldots, [i_s, s, i_{s+1}])$, as a "communication path (of the solution from $[i, r, j]$ to $[k, s, t]$)" if and only if every triplet of the arcs in the set 3-*communicate* in the solution. "In other words, $\{[i_r, r, i_{r+1}], [i_{r+1}, r+1, i_{r+2}], \ldots, [i_s, s, i_{s+1}]\}$ is a "communication path (of (y, z) $[i_r, r, i_{r+1}]$ to $[i_s, s, i_{s+1}]$)" iff $\forall(g, p, q) \in R^3$: $r \leq g < p < q \leq s$, $z_{[i_g,g,i_{g+1}][i_p,p,i_{p+1}][i_q,q,i_{q+1}]} > 0$.

Definition 3.5 ("Spanning Communication Path (SCP)"). We refer to a *communication path* of an LP solution instance, $([i_1, 1, j_1], \ldots, [i_{m-1}, m-1, j_{m-1}])$, as an "SCP (of the solution from $[i_1, 1, j_1]$ to $[i_{m-1}, m-1, j_{m-1}]$)."

The following theorem characterizes *communication paths* in general, and also establishes the equivalence between *spanning communication paths* of an LP solution and *TSP paths* of the TSPFG (and therefore, TSP tours).

Theorem 3.3 (Characterizations of *Communication Paths*).

(1) *The following are true of* communication paths *of an LP solution instance in general*:

    (a) *Every pair of arcs in a given communication path must* 2-*communicate*;

    (b) *For* $(r, s) \in R^2 : s > r$, *arcs* $[i_r, r, j_r], \ldots, [i_s, s, j_s]$ *of the TSPFG can be a communication path of an LP solution only if*:

        (i) $\forall q \in R : r < q \leq s$, $i_q = j_{q-1}$; *and*

        (ii) $\forall p, q \in R : r \leq p < q \leq s$, $i_p \neq i_q \neq j_s$.

(2) *The following are true of spanning communication paths of* $(y, z) \in Q_L$:

    (a) *Every SCP of* (y, z) *is a TSP path of the TSPFG*;

    (b) *Every SCP of* (y, z) *corresponds to exactly one point of* Q_I;

    (c) *Every SCP of* (y, z) *corresponds to exactly one TSP tour*;

    (d) *Every SCP of* (y, z) *corresponds to exactly one extreme point of the LP Polytope (i.e.,* Q_L).

Proof. Statement (1.*a*) follows from the *consistency* constraints (2.16)–(2.18). Statement (1.*b.i*) follows from Definition 3.4. Statement (1.*b.ii*) follows from the combination of Definition 3.4 and the *no-revisit* constraints (2.10). Statement (2.*a*) follows from Definition 2.2 and statement (1.*b*). Statements (2.*b*)–(2.*d*) follow from the combination of statement (1.*a*) and Theorem 2.6. Statement (2.*d*) of the theorem follows from the combination of statement (2.*b*) of the theorem and Theorem 2.7. $\square$

Because of Theorem 3.3, *spanning communication paths* play a key role in our developments. When an LP solution is integral, its support graph (see Notation 3.2 for exact definition) consists of only one *TSP path* of the TSPFG. Hence, Definition 3.5 is sufficient in this case in order to "capture" the corresponding TSP tour. When an LP solution is fractional however, it may be possible to define several SCP's over its support graph. Some of these SCP's may not be meaningful/consistent with respect to the solution, and may be only induced by combinations (such as "patchings"/concatenations of sub-paths, for example) of the meaningful/consistent ones. For example, given a fractional LP solution, say x_1, and a given SCP which is possible over its support graph, if there exists no other point of Q_L, say x_2, such that x_1 can be represented as a convex combination of x_2 and the integral point corresponding to the given SCP, then clearly, the given SCP is inconsistent (not "feasible") with respect to the solution at hand, x_1. Hence, for a fractional LP solution, all of the constraints of Q_L must be taken into account in order to separate the meaningful/consistent SCP's from the non-meaningful/inconsistent ones. This is accomplished through Definition 3.6. In the remainder of this section we will first show that there exists at least one SCP of a given solution which satisfies Definition 3.6. Then, we will use this fact in order to establish the integrality of Q_L.

Definition 3.6 ("Feasible Spanning Communication Path (FSCP)"). Let $(y, z) \in Q_L$. For convenience, write $x = ((y, z))$. Assume the set of SCP's of x is non-empty. Let $\mathcal{P}$ denote a given SCP of x_1, and $\widehat{x}_1$, the characteristic vector of $\mathcal{P}$ (i.e., each positive component of x which is involved in $\mathcal{P}$ is set to 1, and each component of x which is not involved in $\mathcal{P}$ is set to 0 in order to obtain $\widehat{x}_1$). We say that $\mathcal{P}$ is an FSCP of x iff there exist $\varepsilon \in (0, 1]$ and $x_2 \in \mathbb{R}^{\xi_y + \xi_z}$ such that $(\lambda x + (1 - \lambda)x_2) \in Q_L$ for all $\lambda \in [0, 1]$, and $x = \varepsilon\widehat{x}_1 + (1 - \varepsilon)x_2$.

Notation 3.2 ("Support graph" of (y, z)). *For* $(y, z) \in Q_L$:

(1) The sub-graph of the TSPFG induced by the positive components of (y, z) is denoted as:

$$\overline{G}((y, z)) := (\overline{V}((y, z)), \overline{A}((y, z))), \quad \text{where:}$$

$$\overline{V}((y, z)) := \left\{ (i, 1) \in V : \sum_{j \in F_1(i)} \sum_{t \in F_2(j)} y_{[i,1,j][j,2,t]} > 0 \right\}$$

$$\cup \left\{ (i, r) \in V : \left(1 < r < n; \sum_{a \in N_1} \sum_{b \in F_1(a)} \sum_{j \in F_r(i)} y_{[a,1,b][i,r,j]} > 0 \right) \right\}$$

$$\cup \left\{ (i, n) \in V : \sum_{a \in N_1} \sum_{b \in F_1(a)} \sum_{j \in B_n(i)} y_{[a,1,b][j,r-1,i]} > 0 \right\}; \quad (3.7)$$

$$\overline{A}((y, z)) := \left\{ [i, 1, j] \in A : \sum_{t \in F_2(j)} y_{[i,1,j][j,2,t]} > 0 \right\}$$

$$\cup \left\{ [i, r, j] \in A : \left(r > 1; \sum_{a \in N_1} \sum_{b \in F_1(a)} y_{[a,1,b][i,r,j]} > 0 \right) \right\}. \quad (3.8)$$

(2) The set of arcs of $\overline{G}((y, z))$ originating at stage r of $\overline{G}((y, z))$ is denoted $\mathcal{A}_r((y, z))$.

(3) The index set associated with $\mathcal{A}_r((y, z))$ is denoted $\Lambda_r((y, z)) := \{1, 2, \ldots, |\mathcal{A}_r((y, z))|\}$. For simplicity $\Lambda_r((y, z))$ will be henceforth written as Λ_r;

(4) The νth arc in $\mathcal{A}_r((y,z))$ is denoted as $a_{r,\nu}((y,z))$. For simplicity $a_{r,\nu}((y,z))$ will be henceforth written as $a_{r,\nu}$;

(5) For $(r,\nu) \in (R, \Lambda_r)$, the tail of $a_{r,\nu}$ is labeled $\bar{t}_{r,\nu}((y,z))$; the head of $a_{r,\nu}$ is labeled $\bar{h}_{r,\nu}((y,z))$. For simplicity, $\bar{t}_{r,\nu}((y,z))$ will be henceforth written as $\bar{t}_{r,\nu}$, and $\bar{h}_{r,\nu}((y,z))$, as $\bar{h}_{r,\nu}$;

(6) Where that causes no confusion (and where that is convenient), for $(r,s) \in R^2 : s > r$, and $(\rho,\sigma) \in (\Lambda_r, \Lambda_s)$, "$y_{[\bar{t}_{r,\rho},r,\bar{h}_{r,\sigma}][\bar{t}_{s,\sigma},s,\bar{h}_{s,\sigma}]}$" will be henceforth written as "$y_{[r,\rho][s,\sigma]}$". Similarly, for $(r,s,t) \in R^3$ with $r < s < t$ and $(\rho,\sigma,\tau) \in (\Lambda_r, \Lambda_s, \Lambda_t)$, "$z_{[\bar{t}_{r,\rho},r,\bar{h}_{r,\rho}][\bar{t}_{s,\sigma},s,\bar{h}_{s,\sigma}][\bar{t}_{t,\tau},t,\bar{h}_{t,\tau}]}$" will be henceforth written as "$z_{[r,\rho][s,\sigma][t,\tau]}$";

(7) $\forall(r,s) \in R^2 : s \geq r+2$, $\forall(\rho,\sigma) \in (\Lambda_r, \Lambda_s)$:

(a) The set of arcs at stage $(r+1)$ of $\overline{G}((y,z))$ through which flow propagates from $a_{r,\rho}$ onto $a_{s,\sigma}$ is denoted:

$$I_{[r,\rho][s,\sigma]}((y,z)) := \{\lambda \in \Lambda_{r+1} : z_{[r,\rho][r+1,\lambda][s,\sigma]} > 0\}.$$

(b) The set of arcs at stage $(s-1)$ of $\overline{G}((y,z))$ through which flow propagates from $a_{r,\rho}$ onto $a_{s,\sigma}$ is denoted:

$$J_{[r,\rho][s,\sigma]}((y,z)) := \{\mu \in \Lambda_{s-1} : z_{[r,\rho][s-1,\mu][s,\sigma]} > 0\}.$$

(c) The set of all *communication paths* of (y,z) from $a_{r,\rho}$ to $a_{s,\sigma}$ is denoted $\mathcal{W}_{[r,\rho][s,\sigma]}((y,z))$.

(d) The kth element of $\mathcal{W}_{[r,\rho][s,\sigma]}((y,z))$ $(k \in \Pi_{[r,\rho][s,\sigma]}((y,z)))$ is denoted $\mathcal{P}_{[r,\rho],[s,\sigma],k}((y,z))$.

Remark 3.2. *For* $(y,z) \in Q_L$:

(1) An arc of the TSPFG is included in $\overline{G}((y,z))$ if and only if at least one of the flow variables (see Notation 2.2) associated with the arc is positive. Hence, we refer to $\overline{G}((y,z))$ as the "support graph of (y,z)";

(2) $I_{[r,\rho][s,\sigma]}((y,z))$ is the set of arcs at stage $(r+1)$ of the TSPFG that 3-*communicate* (respectively) with arcs $a_{r,\rho}$ and $a_{s,\sigma}$;

(3) $J_{[r,\rho][s,\sigma]}((y,z))$ is the set of arcs at stage $(s-1)$ of the TSPFG that 3-*communicate* (respectively) with arcs $a_{r,\rho}$ and $a_{s,\sigma}$.

(4) It follows directly from the combination of constraints (2.9) and constraints (2.17) that:

$$\forall (r,s) \in R^2 : s \geq r+2, \ \forall (\rho,\sigma) \in (\Lambda_r, \Lambda_s),$$

(i) $y_{[r,\rho][s,\sigma]} > 0 \iff I_{[r,\rho][s,\sigma]}((y,z)) \neq \varnothing;$

(ii) $y_{[r,\rho][s,\sigma]} > 0 \iff J_{[r,\rho][s,\sigma]}((y,z)) \neq \varnothing;$

(iii) $y_{[r,\rho][s,\sigma]} = \displaystyle\sum_{\lambda \in I_{[r,\rho][s,\sigma]}((y,z))} z_{[r,\rho][r+1,\lambda][s,\sigma]}$

$$= \sum_{\mu \in J_{[r,\rho][s,\sigma]}((y,z))} z_{[r,\rho][s-1,\mu][s,\sigma]}.$$

The following lemma is needed in order to prove our main theorem about the "flow" structure of solutions to our LP model.

Lemma 3.1. *Two given arcs $[i,r,j]$ and $[k,s,t]$ (with $s > r$) of the TSPFG 2-communicate in $(y,z) \in Q_L$ if and only if there exists at least one communication path of (y,z) which includes both of the arcs. Moreover, any other (third) arc that 3-communicates with the two given arcs $[i,r,j]$ and $[k,s,t]$ must be part of at least one such communication path of (y,z). In particular, if $[i,r,j]$ and $[k,s,t]$ have a separation equal to zero (i.e., if $s = r+1$), there must exist $[u,r-1,i]$ and $[j,s+1,v]$ (or both), such that $[u,r-1,i]$, $[i,r,j]$, and $[k,s,t]$ 3-communicate, or $[i,r,j]$, $[k,s,t]$, and $[j,s+1,v]$ 3-communicate. If $[i,r,j]$ and $[k,s,t]$ have a separation greater than zero (i.e., if $s > r+1$), there must exist at least one communication path linking them.*

Proof. Let $[a,p,b]$ and $[c,q,d]$ be such that $y_{[a,p,b][c,q,d]} > 0$ in (y,z). We will consider different cases of *Arc Separation* for $[a,p,b]$ and $[c,q,d]$.

Case 1: *Arc Separation* $= 0$ (i.e., $q = p+1$).
From the *no flow break* constraints (2.9), we have:

$$\forall ([i,r,j],[k,r+1,t]) \in A^2, \ y_{[i,r,j][k,r+1,t]} > 0 \implies k = j. \qquad (3.9)$$

From the *consistency* constraints (2.16) and (2.18), we have:

$$\forall \left([i,r,j],[j,r+1,t]\right) \in A^2 : r < m-1, \ y_{[i,r,j][j,r+1,t]} > 0$$

$$\Longleftrightarrow \begin{cases} \text{(i)} & \exists u \in N_{r+3} : z_{[i,r,j][j,r+1,t][t,r+2,u]} > 0 \quad \text{if } r = 1; \\ \text{(ii)} & \exists v \in N_{r-1} : z_{[v,r-1,i][i,r,j][j,r+1,t]} > 0 \quad \text{if } r = m-2; \\ \text{(iii)} & \exists (u,v) \in (N_{r+3}, N_{r-1}) : (z_{[i,r,j][j,r+1,t][t,r+2,u]} > 0 \text{ and} \\ & z_{[v,r-1,i][i,r,j][j,r+1,t]} > 0) \quad \text{if } 1 < r < m-2. \end{cases}$$

$$(3.10)$$

The theorem, in this case, follows directly from the combination of (3.9) and (3.10).

Condition (iii) of (3.10) is illustrated in Figure 3.2. For condition (i) (of (3.10)) which is not shown in the illustration, the stage of the first arc in the pair would be $r = 1$, so that there would be no node to the left of that (first) arc in the TSPFG, so that only the first z-variable illustrated in the figure would be possible. Similarly, for condition (ii) (which also, is not shown), the stage of the second arc in the pair would be $r = m - 1$, so that there would be no node to the right of that (second) arc in the TSPFG, so that only the second z-variable illustrated in the figure would be possible.

Case 2: *Arc Separation* $= 1$ (i.e., $q = p + 2$).

It follows directly from the *consistency* constraints (2.17) that:

$$\forall([i,r,j],[k,r+2,t]) \in A^2 : r \le m-3,$$

$$y_{[i,r,j][k,r+2,t]} > 0 \Longleftrightarrow z_{[i,r,j][j,r+1,k][k,r+2,t]} > 0. \tag{3.11}$$

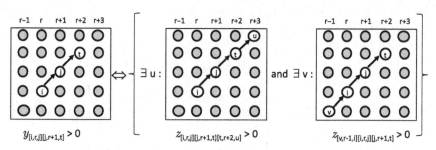

Figure 3.2. Illustration of "Case 1" (i.e., when *arc separation* = 0).

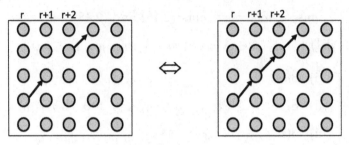

$$y_{[i,r,j][k,r+2,t]} > 0 \iff z_{[i,r,j][j,r+1,t][k,r+2,t]} > 0$$

Figure 3.3. Illustration of "Case 2" (i.e., when *arc separation* = 1).

(In other words, the (unique) arc of the TSPFG, $[j, r + 1, k]$, linking $[i, r, j]$ and $[k, s, t]$ must 3-*communicate* with $[i, r, j]$ and $[k, s, t]$.)

The theorem, in this case, follows from (3.11) directly.

Case 2 is illustrated in Figure 3.3.

Case 3: *Arc Separation* = 2 (i.e., $q = p + 3$).

We will show that $[a, p, b]$ and $[c, q, d]$ can 2-*communicate* in (y, z) if and only if there exists at least one node $((u, p+2) \, (= (u, q-1)))$ of the TSPFG that is such that the arcs $[a, p, b]$, $[b, p+1, u]$, $[u, p+2, c]$, and $[c, q, d]$ form a *communication path* of (y, z), and that, moreover, every arc $[b, p + 1, v]$ that 3-*communicates* with $[a, p, b]$ and $[c, q, d]$ in (y, z) must be part of one such *communication path* of (y, z) from $[a, p, b]$ to $[c, q, d]$. In other words, we will show that the following is true:

$$\forall([i, r, j], [k, r + 3, t]) \in A^2 : r < m - 3,$$

$$y_{[i,r,j][k,r+3,t]} > 0$$

$$\iff \exists u \in N_{r+2} : \begin{cases} \text{(i)} \ \ z_{[i,r,j][j,r+1,u][k,r+3,t]} > 0; \\[2mm] \text{(ii)} \ \ z_{[i,r,j][u,r+2,k][k,r+3,t]} > 0; \\[2mm] \text{(iii)} \ \ z_{[j,r+1,u][u,r+2,k][k,r+3,t]} > 0; \\[2mm] \text{(iv)} \ \ z_{[i,r,j][j,r+1,u][u,r+2,k]} > 0. \end{cases} \quad (3.12)$$

(a) $\Longrightarrow$:

(a.i) *Condition* (i).

$$y_{[i,r,j][k,r+3,t]} = \sum_{a \in N_{r+1}} \sum_{b \in F_{r+1}(a)} z_{[i,r,j][a,r+1,b][k,r+3,t]} \quad \text{(Using (2.17))}$$

$$= \sum_{b \in F_{r+1}(j)} z_{[i,r,j][j,r+1,b][k,r+3,t]} \quad \text{(Using (2.9))} \quad (3.13)$$

$$> 0 \quad \text{(By assumption)}. \quad (3.14)$$

(3.13) and (3.14) $\Longrightarrow \exists u \in N_{r+2} : z_{[i,r,j][j,r+1,u][k,r+3,t]} > 0.$
(a.ii) *Condition* (ii).

Let $u \in N_{r+2}$ be such that it satisfies condition (i) of (3.12). Then, we must have:

$$\sum_{v \in B_{r+2}(u)} z_{[i,r,j][v,r+1,u][k,r+3,t]}$$

$$= z_{[i,r,j][j,r+1,u][k,r+3,t]} \quad \text{(Using (2.9))} \quad (3.15)$$

$$= \sum_{v \in F_{r+2}(u)} z_{[i,r,j][u,r+2,v][k,r+3,t]} \quad \text{(Using (2.3)} \quad (3.16)$$

$$= z_{[i,r,j][u,r+2,k][k,r+3,t]} \quad \text{(Using (2.9))} \quad (3.17)$$

$$> 0 \quad \text{(Using (3.15) and condition } (i)). \quad (3.18)$$

(3.17) and (3.18) $\Longrightarrow z_{[i,r,j][u,r+2,k][k,r+3,t]} > 0.$
(a.iii) *Condition* (iii).

Let $u \in N_{r+2}$ be such that it satisfies conditions (i) and (ii) of (3.12). Using condition (i), and the combination of constraints (2.18) and (2.9), we have:

$$0 < z_{[i,r,j][j,r+1,u][k,r+3,t]} \leq \sum_{a \in B_{r+1}(j)} z_{[a,r,j][j,r+1,u][k,r+3,t]}$$

$$= y_{[j,r+1,u][k,r+3,t]}. \quad (3.19)$$

From Case 2:

$$y_{[j,r+1,u][k,r+3,t]} > 0 \implies z_{[j,r+1,u][u,r+2,k][k,r+3,t]} > 0. \qquad (3.20)$$

Condition (iii) follows directly from the combination of (3.19) and (3.20).

(a.iv) *Condition* (iv).

Let $u \in N_{r+2}$ be such that it satisfies conditions (i)–(iii) of (3.12). Using condition (i), and the combination of constraints (2.16) and (2.9), we have:

$$0 < z_{[j,r+1,u][u,r+2,k][k,r+3,t]} \leq \sum_{b \in F_{r+3}(k)} z_{[i,r,j][u,r+2,k][k,r+3,b]}$$

$$= y_{[u,r+2,k][k,r+3,t]}. \qquad (3.21)$$

From Case 1:

$$y_{[u,r+2,k][k,r+3,t]} > 0 \implies z_{[j,r+1,u][u,r+2,k][k,r+3,t]} > 0. \qquad (3.22)$$

Condition (iv) follows directly from the combination of (3.21) and (3.22).

(b) $\impliedby$:

Follows trivially from the *consistency* constraints (2.17).

Case 3 is illustrated in Figure 3.4. Note that the right-hand-side picture of this figure pertains to only four arcs of the TSPFG, namely, the arcs $[i,r,j]$, $[j,r+1,u]$, $[u,r+2,k]$, and $[k,r+3,t]$. Each line pattern (in the right-hand-side picture) represents a positive z-variable, showing that every combination of three of the four arcs concerned corresponds to a positive z-variable.

Case 4: *Arc Separation* > 2 (i.e., $q > p+3$).

We will prove the theorem for this case by generalizing Cases 2 and 3. For this purpose, it is convenient to use the notation based on the support graph, $\overline{G}((y,z))$, of (y,z). Let $[i,r,j]$ be the ρth arc at stage r of $\overline{G}((y,z))$, and $[k,s,t]$, the σth arc at stage s. Then, we

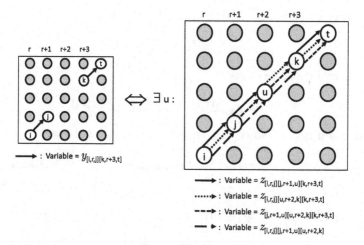

: Variable = $y_{[i,r][j][k,r+3,t]}$

: Variable = $z_{[i,r][j][j,r+1,u][k,r+3,t]}$
: Variable = $z_{[i,r][j][u,r+2,k][k,r+3,t]}$
: Variable = $z_{[j,r+1,u][u,r+2,k][k,r+3,t]}$
: Variable = $z_{[i,r][j][j,r+1,u][u,r+2,k]}$

Figure 3.4. Illustration of "Case 3" (i.e., *arc separation* = 2).

will show that the following statement must hold true:

$$y_{[r,\rho][s,\sigma]} > 0 \Longleftrightarrow \begin{cases} \text{(i) } \mathcal{W}_{[r,\rho][s,\sigma]}((y,z)) \neq \varnothing; \quad \text{and} \\ \\ \text{(ii) } \forall p \in R : r < p < s, \ \forall \nu_p \in \Lambda_p, \\ (z_{[r,\rho][p,\nu_p][s,\sigma]} > 0 \Longleftrightarrow \\ \exists k \in \Pi_{[r,\rho][s,\sigma]}((y,z)) : \\ a_{p,\nu_p} \in \mathcal{P}_{[r,\rho],[s,\sigma],k}((y,z))). \end{cases} \quad (3.23)$$

The proof is by induction. First, note that it follows directly from Cases 2 and 3 that statement (3.23) holds true when the *arc separation*, δ, is equal to 1 or 2. The induction using this is as follows:

(a) $\Longrightarrow$:

Assume there exists an integer $\delta \geq 2$ such that statement (3.23) holds true for all $(r,s) \in R^2$ with $r + 1 \leq s \leq r + \delta + 1$, and all $(\nu_r, \nu_s) \in (\Lambda_r, \Lambda_s)$. We will show that the statement must then also hold for all $(r,s) \in R^2$ with $s = r + \delta + 2$, and all $(\nu_r, \nu_s) \in (\Lambda_r, \Lambda_s)$.

Let $(p,q) \in R^2$ with $q = p + \delta + 2$, and $(\alpha, \beta) \in (\Lambda_p, \Lambda_q)$ be such that:

$$y_{[p,\alpha][q,\beta]} > 0. \quad (3.24)$$

(a.1) From constraints (2.9) and (2.17), we have:

$$I_{[p,\alpha][q,\beta]}((y,z)) \neq \varnothing. \tag{3.25}$$

Constraints (2.18) $\Longrightarrow$

$$\forall \lambda \in I_{[p,\alpha][q,\beta]}((y,z)), \ y_{[p+1,\lambda][q,\beta]} > 0. \tag{3.26}$$

By assumption (since $q = (p+1) + \delta + 1$), (3.26) $\Longrightarrow$

(a.1.1) $\forall \lambda \in I_{[p,\alpha][q,\beta]}((y,z)), \ \mathcal{W}_{[p+1,\lambda][q,\beta]}((y,z)) \neq \varnothing;$ and

$$\tag{3.27a}$$

(a.1.2) $\forall \lambda \in I_{[p,\alpha][q,\beta]}((y,z)), \ \forall t \in R : p+1 < t < q, \ \forall \tau \in \Lambda_t,$

$$(z_{[p+1,\lambda][t,\tau][q,\beta]} > 0 \Longleftrightarrow (\exists i \in \Pi_{[p+1,\lambda][q,\beta]}((y,z)) :$$

$$a_{t,\tau} \in \mathcal{P}_{[p+1,\lambda],[q,\beta],i}((y,z)))). \tag{3.27b}$$

(a.2) From constraints (2.9) and (2.17), we have

$$J_{[p,\alpha][q,\beta]}((y,z)) \neq \varnothing. \tag{3.28}$$

Constraints (2.16) $\Longrightarrow$

$$\forall \mu \in J_{[p,\alpha][q,\beta]}((y,z)), \ y_{[p,\alpha][q-1,\mu]} > 0. \tag{3.29}$$

By assumption (since $(q-1) = p + \delta + 1$), (3.29) $\Longrightarrow$

(a.2.1) $\forall \mu \in J_{[p,\alpha][q,\beta]}((y,z)), \ \mathcal{W}_{[p,\alpha][q-1,\mu]}((y,z)) \neq \varnothing;$ and

$$\tag{3.30a}$$

(a.2.2) $\forall \mu \in J_{[p,\alpha][q,\beta]}((y,z)), \ \forall t \in R : p < t < q-1, \ \forall \tau \in \Lambda_t,$

$$(z_{[p,\alpha][t,\tau][q-1,\mu]} > 0 \Longleftrightarrow (\exists k \in \Pi_{[p,\alpha][q-1,\mu]}((y,z)) :$$

$$a_{t,\tau} \in \mathcal{P}_{[p,\alpha],[q-1,\mu],k}((y,z)))). \tag{3.30b}$$

(a.3) The combination of constraints (2.2)-(2.4) and (2.8) with (3.27a), (3.27b), (3.30a), and (3.30b) $\Longrightarrow$

(a.3.1) $\forall \mu \in \Lambda_{q-1}, \ \exists \langle \lambda \in I_{[p,\alpha][q,\beta]}((y,z)); \ i \in \Pi_{[p+1,\lambda][q,\beta]}((y,z)) \rangle :$

$$\langle a_{q-1,\mu} \in \mathcal{P}_{[p+1,\lambda],[q,\beta],i}((y,z)) \rangle; \text{ and} \tag{3.31a}$$

(a.3.2) $\forall \lambda \in \Lambda_{p+1}, \exists \langle \mu \in J_{[p,\alpha][q,\beta]}((y,z)); \; k \in \Pi_{[p,\alpha][q-1,\mu]}((y,z))\rangle :$

$$\langle a_{p+1,\lambda} \in \mathcal{P}_{[p,\alpha],[q-1,\mu],k}((y,z))\rangle, \tag{3.31b}$$

which, combined with the definitions of $I_{[p,\alpha][q,\beta]}((y,z))$ and $J_{[p,\alpha][q,\beta]}((y,z))$, implies:

$\exists \langle \lambda \in I_{[p,\alpha][q,\beta]}((y,z)); \; i \in \Pi_{[p+1,\lambda][q,\beta]}((y,z)); \; \mu \in J_{[p,\alpha][q,\beta]}((y,z));$

$k \in \Pi_{[p,\alpha][q-1,\mu]}((y,z))\rangle : \langle \forall t \in R : p < t < q, \; \forall \tau \in \Lambda_t :$

$a_{t,\tau} \in \mathcal{P}_{[p+1,\lambda],[q,\beta],i}((y,z)), \; z_{[p,\alpha][t,\tau][q,\beta]} > 0;$

$$\left(\mathcal{P}_{[p+1,\lambda],[q,\beta],i}((y,z))\setminus\{a_{q,\beta}\}\right) = \left(\mathcal{P}_{[p,\alpha],[q-1,\mu],k}((y,z))\setminus\{a_{p,\alpha}\}\right) \neq \varnothing\rangle. \tag{3.32}$$

(a.4) Let $\lambda \in I_{[p,\alpha][q,\beta]}((y,z)),\; i \in \Pi_{[p+1,\lambda][q,\beta]}((y,z)),\; \mu \in J_{[p,\alpha][q,\beta]}((y,z))$, and $k \in \Pi_{[p,\alpha][q-1,\mu]}((y,z))$ be such that they satisfy (3.32). Then, it follows directly from the definitions that

$$\begin{aligned}\overline{P} &:= \{a_{p,\alpha}\} \cup \mathcal{P}_{[p+1,\lambda],[q,\beta],i}((y,z)) \\ &= \{a_{q,\beta}\} \cup \mathcal{P}_{[p,\alpha],[q-1,\mu],k}((y,z)) \end{aligned} \tag{3.33}$$

is a *communication path* of (y,z) from $a_{p,\alpha}$ to $a_{q,\beta}$.

Hence, we have that $\mathcal{W}_{[p,\alpha][q,\beta]}((y,z)) \neq \varnothing$. Also, by assumption, we must have that

$$\forall g \in R : p < g < q, \; \forall \gamma \in \Lambda_g : z_{[p,\alpha][g,\gamma][q,\beta]} > 0,$$

$$\exists i \in \Pi_{[p+1,\lambda][q,\beta]}((y,z)) : a_{g,\gamma} \in \mathcal{P}_{[p,\alpha][q,\beta],i}((y,z)).$$

(since $(p+1 \leq g < p+1+\delta)$ and $(q < g+1+\delta)$.)

(b) $\Longleftarrow$: Follows directly from the definitions and the *consistency* constraints (2.17).

The inductive step of the proof of "Case 4" is illustrated in Figures 3.5 and 3.6.

(c) Conclusion.

Lemma 3.1 follows directly from the combination of Cases 1–4.

$\qquad\qquad\qquad\qquad\qquad\qquad\qquad\qquad\qquad\qquad\qquad\qquad\qquad\qquad$ □

Illustrating: Theorem holds for *Separation* = 2 ⇒ Theorem holds for *Separation* = 3.

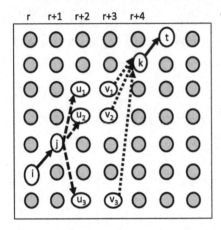

- Let:

  [i, r, j]   (call it **"source-arc"**), and

  [k, r+4, t]   (call it **"destination-arc"**)

  be such that: $y_{[i, r, j][k, r+4, t]} > 0$

- with:

  I = { [j, r+1, u_1], [j, r+1, u_2], [j, r+1, u_3] }
  (call these: **"u-arcs"**) ;

  J = { [v_1, r+3, k], [v_2, r+3, k], [v_3, r+3, k] }
  (call these: **"v-arcs"**).

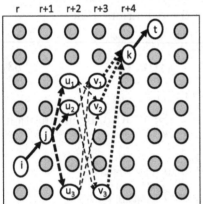

- Then:
1) Each *u-arc* 3-*communicates* (by definition) with the *destination-* and *source-arcs*. Hence, each *u-arc* must 2-*communicate* with the *destination-arc* (because of the *consistency* constraints). Hence, there must exist at least one *communication path* linking each *u-arc* to the *destination-*arc (by assumption, since the *Separation* = 2);
2) Similarly, there must exist at least *one* *communication path* linking each *v-arc* to the *source arc* (since each *v-arc* has *Separation* = 2 from the *source-arc* and must 2-*communicate* with it);
3) Because of the mass balances implicit in the GKE's and the *visit requirements constraints*, at least one pair of *communication paths* (one from the *source-arc* to a *v-arc*, and one from a *u-arc* to the *destination-arc*) must overlap on their portions between the *u-* and *v-arcs* (inclusive).

⇒ Theorem holds for *Separation* = 3.

Figure 3.5. Illustration of the inductive step of "Case 4" when *arc separation* = 3.

Our main result about the "flow" structure of points of Q_L (namely, that if two arcs of the TSPFG 2-*communicate* in a given LP solution, then there must exist at least one FSCP of the solution which includes them both) will now be discussed.

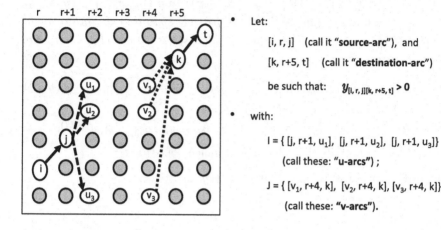

Illustrating: Theorem holds for *Separation* = 3 ⇒ Theorem holds for *Separation* = 4.

- Let:

  [i, r, j] (call it **"source-arc"**), and

  [k, r+5, t] (call it **"destination-arc"**)

  be such that: $y_{[i, r, j][k, r+5, t]} > 0$

- with:

  I = { [j, r+1, u₁], [j, r+1, u₂], [j, r+1, u₃]}
  (call these: **"u-arcs"**) ;

  J = { [v₁, r+4, k], [v₂, r+4, k], [v₃, r+4, k]}
  (call these: **"v-arcs"**).

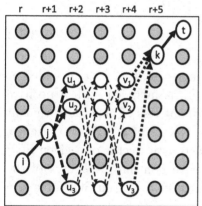

- Then:
1) Each *u-arc* must *3-communicate* with the *destination-* and *source-arcs* (by the *consistency* constraints). Hence, each *u-arc* must *2-communicate* with the *destination-arc* (by the *consistency* constraints). Hence, there must exist at least one *communication path* linking each *u-arc* to the *destination*-arc (by the validity of the theorem when *Separation* = 3);
2) Similarly, there must exist at least *one communication path* linking each *v-arc* to the *source arc* (since each *v-arc* has *Separation* = 3 from the *source-arc* and must *2-communicate* with it);
3) Because of the mass balances implicit in the GKE's and the *visit requirements constraints*, at least one pair of *communication paths* (one from the *source-arc* to a *v-arc*, and one from a *u-arc* to the *destination-arc*) must overlap on their portions between the *u-* and *v-arcs* (inclusive).

⇒ Theorem holds for *Separation* = 4.

Figure 3.6. Illustration of the inductive step of "Case 4" when *arc separation* = 4.

Theorem 3.4. *Two given arcs* $[i, r, j]$ *and* $[k, s, t]$ *(with* $s > r$*) of the TSPFG which 2-communicate in* $(y, z) \in Q_L$ *must be (both) part of at least one FSCP of* (y, z)*. Moreover, any other (third) arc which 3-communicates with the two given arcs* $[i, r, j]$ *and* $[k, s, t]$ *in* (y, z) *must be part of at least one such FSCP of* (y, z)*. In other words, for*

$(y, z) \in Q_L$ *and* $(\alpha, \beta) \in (\Lambda_1, \Lambda_{m-1}) : \mathcal{W}_{[1,\alpha][m-1,\beta]}((y,z)) \neq \varnothing$, *let* $\widehat{\mathcal{W}}_{[1,\alpha][m-1,\beta]}((y,z)) \subseteq \mathcal{W}_{[1,\alpha][m-1,\beta]}((y,z))$ *and* $\widehat{\Pi}_{[1,\alpha][m-1,\beta]}((y,z)) \subseteq \Pi_{[1,\alpha][m-1,\beta]}((y,z))$ *denote the set of feasible spanning communication paths of* (y,z) *and the associated index set, respectively. Then, the following are true:*

(1) $\forall(\alpha, \beta) \in (\Lambda_1, \Lambda_{m-1})$, $y_{[1,\alpha][m-1,\beta]} > 0 \iff (\widehat{\mathcal{W}}_{[1,\alpha][m-1,\beta]}((y,z))$ $\neq \varnothing$, *and* $\widehat{\Pi}_{[1,\alpha][m-1,\beta]}((y,z)) \neq \varnothing)$;

(2) $\forall g \in R : 1 < g < m - 1$, $\forall \gamma \in \Lambda_g : z_{[1,\alpha][g,\gamma][m-1,\beta]} > 0$, $\exists \iota \in$ $\widehat{\Pi}_{[1,\alpha][m-1,\beta]}((y,z)) : a_{g,\gamma} \in \mathcal{P}_{[1,\alpha],[m-1,\beta],\iota}((y,z))$.

Proof. For convenience, write $x = ((y,z))$.

$x \in Q_L \Longrightarrow$:

$$\exists \left\langle \widehat{X} := \{\widehat{x}_1, \ldots, \widehat{x}_p\} \subseteq \text{Ext}(Q_L); (\lambda_1, \ldots, \lambda_p) \in [0,1]^p : \sum_{i=1}^{p} \lambda_i = 1 \right\rangle :$$

$$\left\langle x = \sum_{i=1}^{p} \lambda_i \widehat{x}_i \right\rangle. \tag{3.34}$$

By Theorem 2.7, there exists a representation of x in which at least one integral extreme point of Q_L is included with a positive weight. Hence, it can be assumed without loss of generality that at least one $\widehat{x}_i \in \widehat{X}$ is the characteristic vector of a SCP of x. Hence, it is sufficient to show that the characteristic vector of a given SCP of x can be a non-trivial part (i.e., with non-zero "weight") of the convex combination representation of x iff the given SCP of x is a FSCP of x.

If $p = 1$ in expression (3.34), then it is easy to see that $\widehat{x}_1 = x$ is the characteristic vector of an FSCP of x. Hence, assume $p > 1$ in expression (3.34).

(i) Consider $\widehat{x}_k \in \widehat{X}$ with $0 < \lambda_k < 1$.

Using (3.34), we have:

$$x = \lambda_k \widehat{x}_k + \sum_{i=1; i \neq k}^{p} \lambda_i \widehat{x}_i \tag{3.35}$$

Let $\bar{x} = \sum_{i=1; i \neq k}^{p} \lambda_i \widehat{x}_i$, and $\bar{\bar{x}} = (\frac{1}{1-\lambda_k})\bar{x}$. Then, clearly,

$$\bar{x} \in \widetilde{Q}_L(1 - \lambda_k), \text{ and } \bar{\bar{x}} \in Q_L. \tag{3.36}$$

Hence, (3.35) can be rewitten as:

$$x = \lambda_k \widehat{x}_k + (1 - \lambda_k)\bar{\bar{x}}. \tag{3.37}$$

Also, from (3.37) (since $\bar{\bar{x}} \in Q_L$), we have that:

$$\forall \alpha \in [0, 1], \left(\alpha x + (1 - \alpha)\bar{\bar{x}}\right) \in Q_L. \tag{3.38}$$

It follows from the combination of (3.37) and (3.38) that $\widehat{x}_k$ is the characteristic vector of an FSCP of x.

Hence (by the arbitrariness of $\widehat{x}_k$), if there exists a representation of x in which the characteristic vector of a given SCP of x is included with positive weight, then that SCP of x must be an FCSP of x.

(ii) Suppose $\widetilde{x}$ is the characteristic vector of an FSCP of x. Then, there exist $\varepsilon \in (0, 1]$ and $x_2 \in \mathbb{R}^{\xi_y + \xi_z}$ such that:

$$x = \varepsilon\widetilde{x} + (1 - \varepsilon)x_2, \quad \text{and} \tag{3.39}$$

$$(\lambda x + (1 - \lambda)x_2) \in Q_L \quad \text{for all } \lambda \in (0, 1]. \tag{3.40}$$

Note that each of the constraints in (2.1)–(2.10) is an equality constraint. Hence, the *interior* of Q_L (see Rockafellar (1997, pp. 43–50)) is empty. This combined with (3.40) implies that x_2 must belong to the face of Q_L which includes x. Hence, we must have:

$$\exists \left\langle (\mu_1, \ldots, \mu_p) \in [0, 1]^p : \sum_{i=1}^{p} \mu_i = 1 \right\rangle : \left\langle x_2 = \sum_{i=1}^{p} \mu_i \widehat{x}_i \right\rangle. \tag{3.41}$$

Combining (3.39) with (3.41) gives:

$$x = \varepsilon\widetilde{x} + (1 - \varepsilon)x_2$$

$$= \varepsilon\widetilde{x} + \sum_{i=1}^{p}(1 - \varepsilon)\mu_i \widehat{x}_i. \tag{3.42}$$

For ε and the μ_i's, we have:

$$\varepsilon + \sum_{i=1}^{p}(1-\varepsilon)\mu_i = \varepsilon + \sum_{i=1}^{p}\mu_i - \varepsilon\sum_{i=1}^{p}\mu_i$$

$$= \varepsilon + 1 - \varepsilon$$

$$= 1. \qquad (3.43)$$

Given that $\varepsilon > 0$, statements (3.42) and (3.43), together, imply that $\tilde{x}$ must be included in the representation of x non-trivially. Hence, we must have $\tilde{x} \in \widehat{X}$. $\qquad\qquad\qquad\qquad\qquad\qquad\square$

Remark 3.3. *It follows from (3.35) and (3.36) that if $(\widehat{y}, \widehat{z})$ is the characteristic vector of an FSCP of $(y,z) \in \widetilde{Q}_L(\lambda)$ ($\lambda \in (0,1]$), then:*

(1) $((y,z) - \theta \cdot (\widehat{y}, \widehat{z})) \in \widetilde{Q}_L((\lambda - \theta))$ *for all* $\theta \in (0, \lambda)$; *and*

(2) $(y,z) - \lambda \cdot (\widehat{y}, \widehat{z}) = \mathbf{0}$. (*Since, in this case, the second term of the right-hand-side of (3.35) must be equal to zero, which implies that we must have $(y,z) = \lambda \cdot (\widehat{y}, \widehat{z})$.*)

Definition 3.7. Let $(y,z) \in Q_L$. For every $(\alpha, \beta, i) \in (\Lambda_1, \Lambda_{m-1}, \widehat{\Pi}_{[1,\alpha][m-1,\beta]}((y,z)))$,

(1) we denote the extreme point of Q_L corresponding to the FSCP of (y,z), $\mathcal{P}_{[1,\alpha],[m-1,\beta],i}((y,z))$, by $(\widehat{y}, \widehat{z})^{\alpha\beta i}$ (i.e., $(\widehat{y}, \widehat{z})^{\alpha\beta i}$ is the characteristic vector corresponding to the FSCP of (y,z), $\mathcal{P}_{[1,\alpha],[m-1,\beta],i}((y,z))$);

(2) we say that a y- or z-variable is "included" in $(\widehat{y}, \widehat{z})^{\alpha\beta i}$ if $(\widehat{y}, \widehat{z})^{\alpha\beta i}$ has a "1" entry in its position corresponding to the variable.

4. Integrality of the LP Polytope

In this section, we will show that a feasible solution to the LP must be a convex combination of *TSP paths* (as characterized in terms of the variables).

Definition 3.8. $\forall \lambda \in (0,1]$, $\forall (y,z) \in \widetilde{Q}_L(\lambda)$, $\forall (\alpha, \beta) \in (\Lambda_1, \Lambda_{m-1})$: $\widehat{\mathcal{W}}_{[1,\alpha][m-1,\beta]}((y,z)) \neq \varnothing$, $\forall \iota \in \widehat{\Pi}_{[1,\alpha][m-1,\beta]}((y,z))$, we define:

(1) $\forall \varepsilon \in (0, \lambda]$, $\mathbf{r}_{\lambda,\varepsilon}^{\alpha\beta\iota}((y,z)) := (y,z) - \varepsilon \cdot (\widehat{y}, \widehat{z})^{\alpha\beta\iota}$;

(2) $\delta_\lambda^{\alpha\beta\iota}((y,z)) := \max\left\{\varepsilon \in (0, \lambda] : \mathbf{r}_{\lambda,\varepsilon}^{\alpha\beta\iota}((y,z)) \in \widetilde{Q}_L^0((\lambda - \varepsilon))\right\}$;

(3) $(\overline{y}, \overline{z})_\lambda^{\alpha\beta\iota} := (y,z) - \left(\delta_\lambda^{\alpha\beta\iota}((y,z))\right) \cdot (\widehat{y}, \widehat{z})^{\alpha\beta\iota}$.

For $\lambda \in (0,1]$, $\varepsilon \in (0,\lambda]$, $(y,z) \in \widetilde{Q}_L(\lambda)$, $(\alpha,\beta) \in (\Lambda_1, \Lambda_{m-1})$: $\widehat{\mathcal{W}}_{[1,\alpha][m-1,\beta]}((y,z)) \neq \varnothing$, and $\iota \in \widehat{\Pi}_{[1,\alpha][m-1,\beta]}((y,z))$, $\mathbf{r}^{\alpha\beta\iota}_{\lambda,\varepsilon}((y,z))$ is what is left from $((y,z))$ (the "remainder") after a fraction ε of the extreme point of Q_L, $(\widehat{y},\widehat{z})^{\alpha\beta\iota}$, is subtracted from it; $\delta^{\alpha\beta\iota}_{\lambda}((y,z))$ is the greatest value of ε which is such that the "remainder" $\mathbf{r}^{\alpha\beta\iota}_{\lambda,\varepsilon}((y,z))$ is either feasible for the $(\lambda - \varepsilon)$-*scaled* LP or equal to $\mathbf{0}$; $(\overline{y},\overline{z})^{\alpha\beta\iota}_{\lambda}$ is what is left from the λ-*scaled* LP solution after a fraction $\delta^{\alpha\beta\iota}_{\lambda}((y,z))$ of the extreme point $(\widehat{y},\widehat{z})^{\alpha\beta\iota}$ is subtracted from it.

Theorem 3.5. Q_L *has integral extrema.*

Proof. We will first show that there exists $\varepsilon \in (0,\lambda]$ such that $\mathbf{r}^{\alpha\beta\iota}_{\lambda,\varepsilon}((y,z)) \in \widetilde{Q}^0_L((\lambda - \varepsilon))$. We will then use this to show the integrality of Q_L.

(a) Let $\lambda \in (0,1]$ and $(y,z) \in \widetilde{Q}_L(\lambda)$. Since $\mathcal{P}_{[1,\alpha][m-1,\beta]}((y,z))$ is an FSCP of (y,z), Definition (3.6) $\Longrightarrow$:

$$\exists \langle \theta > 0; (\widetilde{y},\widetilde{z}) \in \mathbb{R}^{\xi_y + \xi_z} \rangle : \langle (y,z) = \theta \cdot (\widehat{y},\widehat{z})^{\alpha\beta\iota} + (1-\theta) \cdot (\widetilde{y},\widetilde{z}) \rangle \tag{3.44}$$

Since $(y,z) \in \widetilde{Q}_L(\lambda)$, (3.44) $\Longrightarrow$:

$$\begin{cases} \theta \leq \lambda, \quad \text{and} \\ (\widetilde{y},\widetilde{z}) \in \widetilde{Q}^0_L(\lambda). \end{cases} \tag{3.45}$$

Using Remark 3.3, (3.44)–(3.45) $\Longrightarrow$:

$$(1-\theta) \cdot (\widetilde{y},\widetilde{z}) = (y,z) - \theta \cdot (\widehat{y},\widehat{z})^{\alpha\beta\iota} = \mathbf{r}^{\alpha\beta\iota}_{\lambda,\theta}((y,z)) \in \widetilde{Q}^0_L((\lambda - \theta)). \tag{3.46}$$

Using Definition 3.8.2, (3.46) $\Longrightarrow$:

$$\delta^{\alpha\beta\iota}_{\lambda}((y,z)) \geq \theta > 0. \tag{3.47}$$

(b) *Convex Hull Representation*

We now show that a solution to the LP must be a convex combination of *TSP paths* (i.e., FSCP's of the solution). In other words, we will show that for $(y,z) \in Q_L$, there must exist $\omega^{\alpha\beta\iota}$'s ($\omega^{\alpha\beta\iota} \in [0,1]$

Table 3.1. Summary of the "IE" procedure.

Step 0 (Initialization)
  (0.1) : **set** $(\overline{y}^0, \overline{z}^0) = (y, z)$
  (0.2) : **set** $X^0 = \{(\alpha, \beta, \iota), (\alpha, \beta) \in (\Lambda_1, \Lambda_{m-1}) :$
$$\widehat{\mathcal{W}}_{[1,\alpha][m-1,\beta]}((y, z)) \neq \varnothing, \ \iota \in \widehat{\Pi}_{[1,\alpha][m-1,\beta]}((y, z))\}$$
  (0.3) : **set** $\mu_0 = 1$
  (0.4) : **set** $k = 0$
Step 1 (Iterative step)
  (1.1) : **set** $k = k + 1$
  (1.2) : **choose** $(\alpha_k, \beta_k, \iota_k) \in X^{k-1}$
  (1.3) : **set** $\theta^k_{\alpha_k, \beta_k, \iota_k} = \delta^{\alpha_k, \beta_k, \iota_k}_{\mu_{k-1}}((\overline{y}^{k-1}, \overline{z}^{k-1}))$
  (1.4) : **compute** $\mu_k = \mu_{k-1} - \theta^k_{\alpha_k, \beta_k, \iota_k}$
  (1.5) : **compute** $(\overline{y}^k, \overline{z}^k) = (\overline{y}^{k-1}, \overline{z}^{k-1}) - \theta^k_{\alpha_k, \beta_k, \iota_k} \cdot (\widehat{y}, \widehat{z})^{\alpha_k, \beta_k, \iota_k}$
  (1.6) : **compute** $X^k = X^{k-1} \setminus \{(\alpha_k, \beta_k, \iota_k)\}$
Step 2 (Stopping criterion)
  (2.1) : **If** $X^k = \varnothing$, **stop**
  (2.2) : **Otherwise, go to** Step 1.

for all $(\alpha, \beta, \iota) \in (\Lambda_1, \Lambda_{m-1}, \widehat{\Pi}_{[1,\alpha][m-1,\beta]}((y, z)))$ such that:

$$\begin{cases} \text{(b.i)} \quad \displaystyle\sum_{\substack{(\alpha,\beta)\in(\Lambda_1,\Lambda_{m-1}): \\ \widehat{\mathcal{W}}_{[1,\alpha][m-1,\beta]}((y,z))\neq\varnothing}} \ \sum_{\iota\in\widehat{\Pi}_{[1,\alpha][m-1,\beta]}((y,z))} \omega^{\alpha\beta\iota} = 1, \quad \text{and} \\[2em] \text{(b.ii)} \quad (y, z) = \displaystyle\sum_{\substack{(\alpha,\beta)\in(\Lambda_1,\Lambda_{m-1}): \\ \widehat{\mathcal{W}}_{[1,\alpha][m-1,\beta]}((y,z))\neq\varnothing}} \ \sum_{\iota\in\widehat{\Pi}_{[1,\alpha][m-1,\beta]}((y,z))} \omega^{\alpha\beta\iota} \cdot (\widehat{y}, \widehat{z})^{\alpha\beta\iota}. \end{cases}$$

(3.48)

The proof of this step is as follows. Let $(y, z) \in Q_L$. Consider the "Iterative Elimination (IE)" procedure shown in Table 3.1. By proof-step (a) above, the *IE procedure* must stop with $(\overline{y}^k, \overline{z}^k) = \mathbf{0}$. Statement (3.48) follows from this directly.

(c) *Integrality of Extreme Points*

From the combination of the *Minkowski–Weyl Theorem* (Minskowski (1910); Weyl (1935); see also Rockafellar (1997, pp. 153–172)) with the proof-step (b) result, we have:

$$\text{Ext}(Q_L) = \{(\widehat{y}, \widehat{z})^{\alpha\beta\iota}, \ (\alpha, \beta) \in (\Lambda_1, \Lambda_{m-1}) : \widehat{\mathcal{W}}_{[1,\alpha][m-1,\beta]}((y, z)) \neq \varnothing,$$

$$\iota \in \widehat{\Pi}_{[1,\alpha][m-1,\beta]}((y, z))\}. \tag{3.49}$$

In light of Definition 3.7, (3.49) and Theorem 3.3 (together) imply:

$$\begin{cases} \text{Ext}(Q_L) = Q_I, & \text{and} \\ Q_L = \text{Conv}(Q_I). \end{cases}$$

$\square$

Corollary 3.1. *It follows directly from Theorem 3.5 that:*

(1) A point of Q_L is an extreme point of Q_L if and only if it consists of exactly one *feasible spanning communication path* (or equivalently, its support graph consists of exactly one *TSP path* of the TSPFG);

(2) There exists a one-to-one correspondence between the extreme points of Q_L and the TSP tours;

(3) Q_L has no integer point in its interior.

Remark 3.4. *"P = NP" has been proved at this point, since it follows directly from Theorem 3.5 and Corollary 3.1 that one can resolve TSP feasibility decision problems (see Garey et al. (1976) for example) by applying arbitrary linear objective cost functions (with appropriate "Big-M" components) over Q_L.*

The resolution of the TSP *optimization problem* requires the development of a linear cost function (in the descriptive variables of Q_L) which will correctly "capture" the TSP tour costs. This is done in Section 5 of this chapter. We will show in the remainder of this section that the set of "weights" stated in (3.48) (proof-step (b) of Theorem 3.5) is unique.

Theorem 3.6. *Consider a solution instance of the LP. The weight (in the extreme point representation of the solution) of any given*

FSCP of the solution that is obtained using the IE procedure is independent of the order in which the given FSCP is chosen.

Proof. Let $\kappa := |\cup_{(\alpha,\beta)\in(\Lambda_1,\Lambda_{m-1})} \widehat{\mathcal{W}}_{[1,\alpha][m-1,\beta]}((y,z))|$, and $K :=$ $\{1,\ldots,\kappa\}$. For $k \in K$, let $\omega^{\alpha_k,\beta_k,\iota_k}$ be the "weight" generated for $\mathcal{P}_{[1,\alpha_k],[m-1,\beta_k],\iota_k}((y,z))$ through the *IE* procedure when $\mathcal{P}_{[1,\alpha_k],[m-1,\beta_k],\iota_k}((y,z))$ is the first FSCP to be chosen in the procedure. To simplify the exposition, write the $(\widehat{y},\widehat{z})^{\alpha\beta\iota}$ $((\alpha,\beta) \in (\Lambda_1,\Lambda_{m-1}) : \widehat{\mathcal{W}}_{[1,\alpha][m-1,\beta]}((y,z)) \neq \varnothing, \iota \in \widehat{\Pi}_{[1,\alpha][m-1,\beta]}((y,z)))$ as $\widehat{x}_1,\ldots,\widehat{x}_\kappa$. Let the corresponding $\omega^{\alpha_k,\beta_k,\iota_k}$'s be written as $w_1,\ldots,w_\kappa$, respectively.

Clearly, the theorem holds trivially if $\kappa = 1$. Hence, assume $\kappa > 1$. To prove the theorem, we will consider two runs of the *IE procedure* in which the first extreme points to be chosen are different. Let $\widehat{x}_p$ $(1 \leq p \leq \kappa)$ be the extreme point chosen first in the first run of the procedure (call it "run #1"), and let $\widehat{x}_q$ $(1 \leq q \leq \kappa; q \neq p)$ be the extreme point chosen first in the second run of the procedure (call it "run #2").

First, consider "run #1". Writing the weights generated by this run (i.e., the $\theta^k_{\alpha_k,\beta_k,\iota_k}$ $(k = 1,\ldots,\kappa)$) as $\lambda^{(1)}_k$ $(k = 1,\ldots,\kappa)$, we must have:

$$\begin{cases} \text{(i)} & \lambda^{(1)}_p = w_p; \\ \text{(ii)} & \lambda^{(1)}_i \leq w_i, \ i \in K\backslash\{p\}; \\ \text{(iii)} & w_p + \lambda^{(1)}_q + \sum_{i\in K\backslash\{p,q\}} \lambda^{(1)}_i = 1; \\ \text{(iv)} & (y,z) = w_p\cdot\widehat{x}_p + \lambda^{(1)}_q\cdot\widehat{x}_q + \sum_{i\in K\backslash\{p,q\}} \lambda^{(1)}_i\cdot\widehat{x}_i. \end{cases} \quad (3.50)$$

Second, consider "run #2" of the *IE procedure*. Let the generated weights be $\lambda^{(2)}_1, \lambda^{(2)}_2,\ldots,\lambda^{(2)}_\kappa$ for $\widehat{x}_1,\ldots,\widehat{x}_\kappa$, respectively. Then, we must have:

$$
\begin{cases}
\text{(i)} \ \ \lambda_q^{(2)} = w_q; \\[2mm]
\text{(ii)} \ \ \lambda_i^{(2)} \leq w_i, \ \ i \in K \backslash \{q\}; \\[2mm]
\text{(iii)} \ \ \lambda_p^{(2)} + w_q + \sum_{i \in K \backslash \{p,q\}} \lambda_i^{(2)} = 1; \\[4mm]
\text{(iv)} \ \ (y,z) = \lambda_p^{(2)} \cdot \widehat{x}_p + w_q \cdot \widehat{x}_q + \sum_{i \in K \backslash \{p,q\}} \lambda_i^{(2)} \cdot \widehat{x}_i.
\end{cases} \tag{3.51}
$$

Thirdly, subtract (3.51.iv) from (3.50.iv). This gives:

$$
(w_p - \lambda_p^{(2)}) \cdot \widehat{x}_p + (\lambda_q^{(1)} - w_q) \cdot \widehat{x}_q + \sum_{i \in K \backslash \{p,q\}} (\lambda_i^{(1)} - \lambda_i^{(2)}) \cdot \widehat{x}_i = 0. \tag{3.52}
$$

Now, (3.52) and Theorem 3.1 $\Longrightarrow$

$$
\begin{cases}
\text{(i)} \ \ \lambda_p^{(2)} = w_p; \\[2mm]
\text{(ii)} \ \ \lambda_q^{(1)} = w_q; \\[2mm]
\text{(iii)} \ \ \lambda_i^{(1)} = \lambda_i^{(2)}, \ i \in K \backslash \{p,q\}.
\end{cases} \tag{3.53}
$$

The theorem follows the combination of (3.50.i), (3.51.i), (3.53), and the fact that the weights produced by the *IE procedure* based on any (arbitrary) order of $\widehat{x}_1, \ldots, \widehat{x}_\kappa$ are valid for the convex combination representation of (y,z) in terms of $\widehat{x}_1, \ldots, \widehat{x}_\kappa$. $\qquad \square$

Theorem 3.7. *For* $(y,z) \in Q_L$, *let* $(\widehat{y}, \widehat{z})^{\alpha\beta\iota}$ *and* $\omega^{\alpha\beta\iota}$ *denote the characteristic vector and "weight" obtained for the FSCP* $\mathcal{P}_{[1,\alpha],[m-1,\beta],\iota}((y,z))$ *using the IE procedure, respectively. Then, the following are true:*

(1) $\{(\widehat{y}, \widehat{z})^{\alpha\beta\iota} : (\alpha,\beta) \in (\Lambda_1, \Lambda_{m-1}), \widehat{\mathcal{W}}_{[1,\alpha][m-1,\beta]}((y,z)) \neq \varnothing, \iota \in \widehat{\Pi}_{[1,\alpha][m-1,\beta]}((y,z))\}$ *is affinely independent;*

(2) $\{\omega^{\alpha\beta\iota} \; : \; (\alpha,\beta) \in (\Lambda_1, \Lambda_{m-1}), \widehat{\mathcal{W}}_{[1,\alpha][m-1,\beta]}((y,z)) \neq \varnothing, \; \iota \in \widehat{\Pi}_{[1,\alpha][m-1,\beta]}((y,z))\}$ *is a barycentric coordinate system for* (y,z).

Proof. The theorem follows directly from the combination of the result of proof-step (c) of Theorem 3.5 (i.e., (3.48)) and Theorem 3.6 (see Rockafellar (1997, pp. 3–9) or Magaril-Il'yaev and Tikhomirov (2000, pp. 25–33), among others). $\qquad\square$

Corollary 3.2. *Let* $(y,z) \in Q_L$, $(\widehat{y},\widehat{z})^{\alpha\beta\iota}$, *and* $\omega^{\alpha\beta\iota}$ *denote the characteristic vector and "weight" obtained for the FSCP* $\mathcal{P}_{[1,\alpha],[m-1,\beta],\iota}((y,z))$ *using the IE procedure. Then, the following are true:*

(1) $\forall (r,s) \in R^2 : r < s, \; \forall(\nu_r, \nu_s) \in (\Lambda_r, \Lambda_s)$,

$$y_{[r,\nu_r][s,\nu_s]} = \sum_{\alpha\in\Lambda_1} \sum_{\beta\in\Lambda_{n-1}} \sum_{\substack{\iota\in\widehat{\Pi}_{[1,\alpha][n-1,\beta]}((y,z)): \\ (a_{r,\nu_r},\, a_{s,\nu_s})\in\mathcal{P}^2_{[1,\alpha],[m-1,\beta],\iota}((y,z))}} \omega^{\alpha\beta\iota};$$

(2) $\forall (r,s,t) \in R^3 : r < s < t, \; \forall(\nu_r, \nu_s, \nu_t) \in (\Lambda_r, \Lambda_s, \Lambda_t)$,

$$z_{[r,\nu_r][s,\nu_s][t,\nu_t]} = \sum_{\alpha\in\Lambda_1} \sum_{\beta\in\Lambda_{n-1}} \sum_{\substack{\iota\in\widehat{\Pi}_{[1,\alpha][m-1,\beta]}((y,z)): \\ (a_{r,\nu_r},\, a_{s,\nu_s},\, a_{t,\nu_t})\in\mathcal{P}^3_{[1,\alpha],[m-1,\beta],\iota}((y,z))}} \omega^{\alpha\beta\iota}.$$

Corollary 3.2 is illustrated in the following examples. In Example 3.1, a trivial case is shown in which the FSCP's do not overlap. Example 3.2 is a more general case, but each FSCP has at least one positive y-variable and at least one positive z-variable that belong to it to the exclusion of all other FSCP's. Example 3.3 shows a case where each positive y-variable and each positive z-variable belongs to more than one FSCP, respectively.

Example 3.1. *Illustration of Corollary 3.2 for the special (simpliest) case of $n = 5$:*

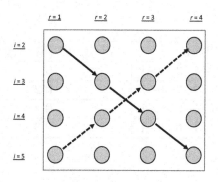

➝ : Example fractional path; flow = 1/2

⇢ : Example fractional path; flow = 1/2

□

Example 3.2. *Illustration of Corollary 3.2 for the general (i.e., $n > 5$) case in which each FSCP/TSP path has a (y- and z-) variable that belongs to it exclusively:*

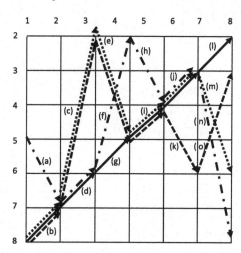

- Let "initial" z's be:
  - $z_{[a][d][f]} = 0.1$,
  - $z_{[b][c][e]} = 0.5$,

— (from KE's, let: $z_{[b][c][k]} = 0.3$; $z_{[b][c][j]} = 0.2$),

— $z_{[b][d][g]} = 0.4$.

- TSP "paths"

TSP Tour	"Weight" ($\delta_1^{\alpha\beta\iota}((\mathbf{y},\mathbf{z}))$'s)
1–5–7–6–2–4–3–8–1	0.1
1–8–7–2–5–4–3–6–1	0.2
1–8–7–6–5–4–3–2–1	0.4
1–8–7–2–5–4–6–3–1	0.3

- $z_{[b][i][j]} = 0.6$?

  — $z_{[b][i][j]} = z_{[b][e][j]} + z_{[b][g][j]}$ (KE: (b), (j), $(5,4)$),

  — $z_{[b][e][j]} = z_{[b][c][j]} = 0.2$,

  — $z_{[b][g][j]} = z_{[b][d][j]} = z_{[b][d][i]} = z_{[b][d][g]} = 0.4$.

  — $\Longrightarrow$Yes!

- $y_{[b][d]} = 0.4$?

  — $y_{[b][d]} = z_{[b][d][g]} = 0.4$.

  — $\Longrightarrow$Yes!

- $y_{[b][i]} = 0.9$?

  — $y_{[b][i]} = z_{[b][e][i]} + z_{[b][g][i]}$,

  — $z_{[b][e][i]} = z_{[b][c][i]} = z_{[b][c][e]} = 0.5$,

  — $z_{[b][g][i]} = z_{[b][d][i]} = z_{[b][d][g]} = 0.4$.

  — $\Longrightarrow$Yes!

- $y_{[d][j]} = 0.5$?

  — $y_{[d][j]} = z_{[a][d][j]} + z_{[b][d][j]}$,

  — $z_{[a][d][j]} = z_{[a][d][h]} = z_{[a][d][f]} = 0.1$,

  — $z_{[b][d][j]} = z_{[b][d][i]} = z_{[b][d][g]} = 0.4$.

  — $\Longrightarrow$Yes!

- Etc.

Example 3.3. *Illustration of Corollary 3.2 for the general (i.e., $n > 5$) case in which each (y- and z-) variable belongs to more than one FSCP/TSP path. This is the case in which every pair of arcs of the support graph that 2-communicate in a solution belongs to at least two FSCP's of the solution, and every triplet of arcs of the support graph which 3-communicate in the solution belongs to at least two FSCP's of the solution. First, we provide a "construct" for the general structure of such a solution. Then, we illustrate the application of Corollary 3.2 to this solution numerically.*

- Support graph:

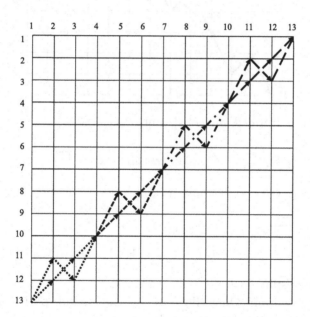

— An FSCP is obtained by picking one sub-path from each line style.

— Hence, the total number of FSCP's is 16.

— Observation: Each triplet of arcs obtained through the process above belongs to more than one FSCP.

- **Case 1**: Illustration of the flow propagation along an FSCP when all the FSCP's have equal "weights":

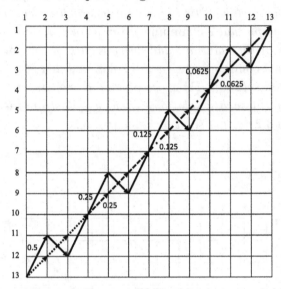

— By the *Initial Flow* constraint (2.1), let

$$z_{[13,1,11][11,2,12][12,3,10]} = 0.5;$$

— By the *GKE* constraints (2.2):

$$z_{[13,1,11][11,2,12][12,3,10]}$$
$$= z_{[13,1,11][11,2,12][10,4,8]} + z_{[13,1,11][11,2,12][10,4,9]}$$

(The "split" will be $(0.25, 0.25)$);
— Observation: Once flow along the "solid path" splits (as happens at node $(10, 4)$ above), it cannot subsequently "re-combine" (at node $(7, 7)$ for example; according to the GKE's) with the flow on the "dashed path";
— The "weight" of the "solid path" in the solution is 0.0625 (as computed using Corollary 3.2).

- **Case 2**: Illustration of the flow propagation along an FSCP when the FSCP's have unequal "weights":

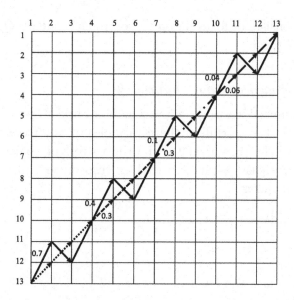

— By the *Initial Flow* constraint 2.1, let

$$z_{[13,1,11][11,2,12][12,3,10]} = 0.7;$$

— By the *GKE* constraints 2.2:

$$z_{[13,1,11][11,2,12][12,3,10]}$$
$$= z_{[13,1,11][11,2,12][10,4,8]} + z_{[13,1,11][11,2,12][10,4,9]}$$

(Let it be the "split" as shown);

— Observation: Once flow along the "solid path" splits (as happens at node $(10, 4)$ above), it cannot subsequently "re-combine" (at node $(7, 7)$ for example; according to the GKE's) with the flow on the "dashed path".

— The "weight" of the "solid path" in the solution with the "splits" shown is 0.04 (as computed using Corollary 3.2). (Any other set of "splits" corresponds to a different LP solution.)

5. Linear Cost Function for the *TSP Paths*

In this section we develop alternate linear cost functions for the TSP tours in the variables used to describe our LP polytope, Q_L. The approach consists of focusing on the extreme points of Q_L (i.e., the points of Q_I).

5.1. *"Balanced costing" approach*

In the "balanced costing" approach, non-implicitly-zero costs are associated with either all of the y-variables or all of the z-variables of the model. Letting $\mathfrak{b}^y$ and $\mathfrak{b}^z$ be the vectors of "balanced costs" associated with the y- and z-variables respectively, these costs are specified as follows:

$$\forall([i,r,j],\,[k,s,t])\in A^2 : r < s,$$

$$\mathfrak{b}^y_{[i,r,j][k,s,t]}$$
$$:= \begin{cases} (c_{1,i} + c_{i,j} + c_{k,t})\,/\,(m-2) & \text{if } (r=1;\ s < m-1); \\ (c_{1,i} + c_{i,j} + c_{k,t} + c_{t,1})\,/\,(m-2) & \text{if } (r=1;\ s = m-1); \\ (c_{i,j} + c_{k,t})\,/\,(m-2) & \text{if } (r>1;\ s < m-1); \\ (c_{i,j} + c_{k,t} + c_{t,1})\,/\,(m-2) & \text{if } (r>1;\ s = m-1); \\ 0 & \text{otherwise,} \end{cases}$$

and

$$\forall([i,r,j],\,[k,s,t],\,[u,p,v])\in A^3 : r < s < p,$$

$$\mathfrak{b}^z_{[i,r,j][k,s,t][u,p,v]}$$
$$:= \begin{cases} (2*(c_{1,i} + c_{i,j} + c_{k,t} + c_{u,v}))\,/\,((m-2)\,(m-3)), \\ \quad \text{if } (r=1;\ p < m-1); \\ (2*(c_{1,i} + c_{i,j} + c_{k,t} + c_{u,v} + c_{v,1}))\,/\,((m-2)\,(m-3)), \\ \quad \text{if } (r=1;\ p = m-1); \\ (2*(c_{i,j} + c_{k,t} + c_{u,v}))/\,((m-2)\,(m-3)), \\ \quad \text{if } (r>1;\ p < m-1); \\ (2*(c_{i,j} + c_{k,t} + c_{u,v} + c_{t,1}))/\,((m-2)\,(m-3)), \\ \quad \text{if } (r>1;\ p = m-1); \\ 0 \quad \text{otherwise,} \end{cases}$$

Example 3.4 (Illustration of the "balanced costs").

- *Inter-city travel costs, $c_{i,j}$'s*:

	1	2	3	4	5	6	7
1	–	44	13	8	47	27	36
2	36	–	35	40	25	19	26
3	5	17	–	17	32	35	38
4	30	5	19	–	12	46	29
5	31	2	28	33	–	46	7
6	26	39	27	14	49	–	12
7	11	37	20	36	23	27	–

- *TSP tour*:

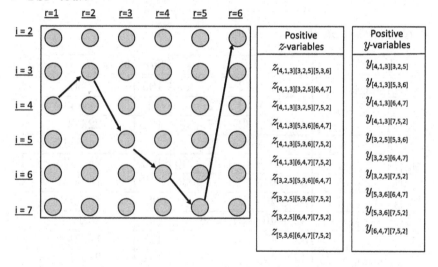

Positive z-variables
$z_{[4,1,3][3,2,5][5,3,6]}$
$z_{[4,1,3][3,2,5][6,4,7]}$
$z_{[4,1,3][3,2,5][7,5,2]}$
$z_{[4,1,3][5,3,6][6,4,7]}$
$z_{[4,1,3][5,3,6][7,5,2]}$
$z_{[4,1,3][6,4,7][7,5,2]}$
$z_{[3,2,5][5,3,6][6,4,7]}$
$z_{[3,2,5][5,3,6][7,5,2]}$
$z_{[3,2,5][6,4,7][7,5,2]}$
$z_{[5,3,6][6,4,7][7,5,2]}$

Positive y-variables
$y_{[4,1,3][3,2,5]}$
$y_{[4,1,3][5,3,6]}$
$y_{[4,1,3][6,4,7]}$
$y_{[4,1,3][7,5,2]}$
$y_{[3,2,5][5,3,6]}$
$y_{[3,2,5][6,4,7]}$
$y_{[3,2,5][7,5,2]}$
$y_{[5,3,6][6,4,7]}$
$y_{[5,3,6][7,5,2]}$
$y_{[6,4,7][7,5,2]}$

- *Costs based on the y-variables*:

Variable, y	Terms	Sum	$\mathfrak{b}^y = Sum/4$
$y_{[4,1,3][3,2,5]}$	$8 + 19 + 32$	59	14.7500
$y_{[4,1,3][5,3,6]}$	$8 + 19 + 46$	73	18.2500
$y_{[4,1,3][6,4,7]}$	$8 + 19 + 12$	39	9.7500
$y_{[4,1,3][7,5,2]}$	$8 + 19 + 37 + 36$	100	25.0000
$y_{[3,2,5][5,3,6]}$	$32 + 46$	78	19.5000
$y_{[3,2,5][6,4,7]}$	$32 + 12$	44	11.0000
$y_{[3,2,5][7,5,2]}$	$32 + 37 + 36$	105	26.2500
$y_{[5,3,6][6,4,7]}$	$46 + 12$	58	14.5000
$y_{[5,3,6][7,5,2]}$	$46 + 37 + 36$	119	29.7500
$y_{[6,4,7][7,5,2]}$	$12 + 37 + 36$	85	21.25000
$Total =$	–	760	190.0000

- *Costs based on the z-variables*:

Variable, z	Terms	Sum	$\mathfrak{b}^z = 2 * Sum/12$
$z_{[4,1,3][3,2,5][5,3,6]}$	$8 + 19 + 32 + 46$	105	17.5000
$z_{[4,1,3][3,2,5][6,4,7]}$	$8 + 19 + 32 + 12$	71	11.8333
$z_{[4,1,3][3,2,5][7,5,2]}$	$8 + 19 + 32 + 37 + 36$	132	22.0000
$z_{[4,1,3][5,3,6][6,4,7]}$	$8 + 19 + 46 + 12$	85	14.1667
$z_{[4,1,3][5,3,6][7,5,2]}$	$8 + 19 + 46 + 37 + 36$	146	24.3333
$z_{[4,1,3][6,4,7][7,5,2]}$	$8 + 19 + 12 + 37 + 36$	112	18.6667
$z_{[3,2,5][5,3,6][6,4,7]}$	$32 + 46 + 12$	90	15.0000
$z_{[3,2,5][5,3,6][7,5,2]}$	$32 + 46 + 37 + 36$	151	25.1667
$z_{[3,2,5][6,4,7][7,5,2]}$	$32 + 12 + 37 + 36$	117	19.5000
$z_{[5,3,6][6,4,7][7,5,2]}$	$46 + 12 + 37 + 36$	131	21.8333
$Total =$	–	1140	190.0000

□

5.2. *"Un-balanced costing" approach*

In the "un-balanced costing" approach, the costs associated with most of the variables are set to zero. Only a relatively small subset of the variables have costs that are not set to zero. Letting $\mathfrak{u}^y$ and

u^z be the vectors of "un-balanced costs" associated with the y- and z-variables, respectively, these costs are specified as follows:

$$\forall ([i,r,j],\ [k,s,t]) \in A^2 : r < s,$$

$$u^y_{[i,r,j][k,s,t]} := \begin{cases} c_{1,i} + c_{i,j} + c_{j,t} & \text{if } (r = 1;\ s = 2;\ k = j); \\ c_{k,t} & \text{if } (r = 1;\ 2 < s < m - 1); \\ c_{k,t} + c_{t,1} & \text{if } (r = 1;\ s = m - 1); \\ 0 & \text{otherwise,} \end{cases}$$

and,

$$\forall ([i,r,j],\ [k,s,t],\ [u,p,v]) \in A^3 : r < s < p,\ \text{let}$$

$$u^z_{[i,r,j][k,s,t][u,p,v]}$$

$$:= \begin{cases} c_{1,i} + c_{i,j} + c_{j,t} + c_{t,v} & \text{if } (r = 1;\ s = 2;\ p = 3;\ k = j;\ u = t); \\ c_{u,v} & \text{if } (r = 1;\ s = 2;\ 3 < p < m - 1;\ k = j); \\ c_{u,v} + c_{v,1} & \text{if } (r = 1;\ s = 2;\ p = m - 1;\ k = j); \\ 0 & \text{otherwise.} \end{cases}$$

Example 3.5 (Illustration of the "un-balanced costs"). *Referring to the TSP of Example 3.4, the "un-balanced costs" are as follows*:

- *Costs based on the y-variables:*

Variable, y	Terms	$u^y = Sum$
$y_{[4,1,3][3,2,5]}$	$8 + 19 + 32$	59
$y_{[4,1,3][5,3,6]}$	46	46
$y_{[4,1,3][6,4,7]}$	12	12
$y_{[4,1,3][7,5,2]}$	$37 + 36$	73
$y_{[3,2,5][5,3,6]}$	0	0
$y_{[3,2,5][6,4,7]}$	0	0
$y_{[3,2,5][7,5,2]}$	0	0
$y_{[5,3,6][6,4,7]}$	0	0
$y_{[5,3,6][7,5,2]}$	0	0
$y_{[6,4,7][7,5,2]}$	0	0
$Total =$	–	190

• *Costs based on the z-variables:*

Variable, z	Terms	$u^z = sum$
$z_{[4,1,3][3,2,5][5,3,6]}$	$8 + 19 + 32 + 46$	105
$z_{[4,1,3][3,2,5][6,4,7]}$	12	12
$z_{[4,1,3][3,2,5][7,5,2]}$	$37 + 36$	73
$z_{[4,1,3][5,3,6][6,4,7]}$	0	0
$z_{[4,1,3][5,3,6][7,5,2]}$	0	0
$z_{[4,1,3][6,4,7][7,5,2]}$	0	0
$z_{[3,2,5][5,3,6][6,4,7]}$	0	0
$z_{[3,2,5][5,3,6][7,5,2]}$	0	0
$z_{[3,2,5][6,4,7][7,5,2]}$	0	0
$z_{[5,3,6][6,4,7][7,5,2]}$	0	0
$Total =$	–	190

□

5.3. Alternate objective functions for the overall LP model

Theorem 3.8. *Let* $(y, z) \in \text{Ext}(Q_L)$. *Let* $\lambda_1, \lambda_2, \lambda_3, \lambda_4$ *be scalars on the interval* $[0, 1]$ *with* $\sum_{i=1}^{4} \lambda_i = 1$. *Then, the following linear function in the y- and z-variables:*

$$\vartheta(y, z) := \left(\lambda_1 b^y + \lambda_2 u^y\right)^T \cdot y + \left(\lambda_3 b^z + \lambda_4 u^z\right)^T \cdot z$$

accurately accounts the cost of the TSP tour corresponding to (y, z).

Proof. First, note that Theorems 2.8 and 3.5 $\implies$ $((y, z) \in \text{Ext}(Q_L) \implies (y, z) \in Q_I)$. It is trivial to verify that each of $(b^y, \mathbf{0})$, $(\mathbf{0}, b^z)$, $(u^y, \mathbf{0})$, and $(\mathbf{0}, u^z)$ accurately accounts the cost of the TSP tour corresponding to (y, z). The theorem follows from this directly. □

In our computational testing we have tried both balanced and un-balanced objectives as well as convex combinations of balanced (y-variables and z-variables cases) and unbalanced (y-variables and z-variables cases). They have all worked with the combined balanced case often solving more quickly.

Chapter 4

Generic LP Modeling for COPs

1. Introduction

In this chapter, we present a unifying perspective on combinatorial optimization problems (COPs) that allows for the straightforward extension of the model developed in Chapters 2 and 3 to other hard COPs. First, we provide a new, unifying description of COPs. Then, we develop a Bipartite Network Flow (BNF) Problem-based formulation that is generic for many of the well-known COPs. For a given COP, using this BNF-based formulation, we develop the flow graph and a path representation of solutions, in a similar way as for the TSP (as described in Chapter 2). After the flow graph of a given COP has been formalized, our proposed model (as stated in Chapter 2 and developed in Chapter 3 of this book) can be directly written (in terms of the nodes and their backward and forward stars) for the specific COP at hand.

2. Unified Description and Classification of COPs

The unifying, generic description of COPs we propose is as follows. There is a ground set $\mathbb{O} = \{o_1, \ldots, o_u\}$ of objects, and a family of subsets, $\{\mathbb{F}_j\}_{j \in \mathbb{J} := \{1, \ldots, s\}}$, of this ground set representing grouping/assignment possibilities for the objects. The $\mathbb{F}_j's$ $(j \in \mathbb{J})$ can be viewed from one of two broad perspectives that are typified by the contexts of crew scheduling and facility location, respectively, based on natural restrictions which exist in one (crew scheduling) and not in the other (location). In a crew scheduling context (such as that of surgical theater planning) each $\mathbb{F}_j$ $(j \in \mathbb{J})$ would normally be one "unbreakable" unit (a "crew"), so that it would not be physically possible for a given object to be assigned to more than one of the $\mathbb{F}_j$'s. On the other hand, in the context of locating hospitals for example, a given population center (the "objects") can be assigned to more than one location site (the $\mathbb{F}_j$'s), and not every population center that is assignable to a given location site need be assigned to the site when the site is active. We refer to the $\mathbb{F}'_j s$ that are subject to the restrictions of crew scheduling contexts described above as "crew subsets," and to the $\mathbb{F}'_j s$ that are not subject to these restrictions as "free subsets".

Example 4.1.

Let:

- $\mathbb{O} = \{1, 2, 3, 4, 5\}$;
- $\mathfrak{s} = 5$;
- $\mathbb{F}_1 = \{1, 2\}$,
  $\mathbb{F}_2 = \{2, 3\}$,
  $\mathbb{F}_3 = \{1, 3, 4\}$,
  $\mathbb{F}_4 = \{2, 4, 5\}$,
  $\mathbb{F}_5 = \{1, 2, 3, 4, 5\}$.

Then, for example:

- *The "assignment possibilities to" $\mathbb{F}_1$ are (objects) "1" and "2".*
- *The "assignment possibilities of" object "1" are $\mathbb{F}_1$, $\mathbb{F}_3$, and $\mathbb{F}_5$.*
- *If $\mathbb{F}_1$ were a crew subset, then there would be a restriction that either both objects "1" and "2" are assigned to $\mathbb{F}_1$ or neither of them is.*
- *If $\mathbb{F}_1$ were a free subset, then it would be feasible to assign to it exactly one of objects "1" or "2," or both objects "1" and "2," or neither of the objects.*

To each $\mathbb{F}_j$ $(j \in \mathbb{J})$ we associate a "fixed cost", $\mathfrak{f}_j$, and sequence-dependent costs, $\mathfrak{d}_{j,i,k}$ $((i, k) \in \mathbb{F}_j^2)$, which are incurred if objects i and k are assigned to $\mathbb{F}_j$ in order, with k immediately following i. The

$\mathfrak{f}_j$'s and $\mathfrak{d}_{j,i,k}$'s are not restricted in sign in our proposed framework. Hence, we are using the term "cost" in a broad, generic sense.

Example 4.2. *In reference to Example 4.1:*

Let the costs associated with $\mathbb{F}_3$ *be as follows:*

- $\mathfrak{f}_3 = 100$
- $\mathfrak{d}_{3,i,k}$ *'s:*

	$k = 1$	$k = 3$	$k = 4$
$i = 1$	—	40	20
$i = 3$	10	—	50
$i = 4$	50	30	—

Then:

- $\mathbb{F}_3$ *is said to be "chosen" if objects in any subset of* $\{1, 3, 4\}$ *are assigned to it;*
- *If objects "1" and "3" are assigned to* $\mathbb{F}_3$ *in the order "1"* $\longrightarrow$ *"3", the total cost incurred for this choice will be:* $100 + 40 = 140$;
- *If objects "1" and "3" are assigned to* $\mathbb{F}_3$ *in the order "3"* $\longrightarrow$ *"1", the total cost incurred for this choice will be:* $100 + 10 = 110$;
- *Similarly, if all three objects "1," "2", and "3" are assigned to* $\mathbb{F}_1$, *the total cost incurred will depend on the sequence in which the assignments are made.*

The objective of our optimization problem is to make assignments of the objects in $\mathbb{O}$ to the $\mathbb{F}'_j s$ ($j \in \mathbb{J}$) so as to minimize the total fixed and sequence-dependent costs. We say that a given $\mathbb{F}_j$ ($j \in \mathbb{J}$) is "chosen" ("not chosen") in a given solution instance of the optimization problem if one or more (none) of the objects in $\mathbb{O}$ is assigned to it. If a given $\mathbb{F}_j$ ($j \in \mathbb{J}$) is chosen, then the total cost incurred is $\mathfrak{f}_j$ plus the total of the sequence-dependent costs of the specific permutation chosen for the subset of objects assigned to $\mathbb{F}_j$. On the other hand, if a given $\mathbb{F}_j$ ($j \in \mathbb{J}$) is not chosen, then the total cost incurred in the solution by that $\mathbb{F}_j$ is zero.

The constraints of the optimization problem are the "coverage", "mutual exclusivity", and/or "joint inclusion" constraints. The coverage constraints are requirements that each object in $\mathbb{O}$ be assigned to exactly (or at least) one of the $\mathbb{F}_j s$ ($j \in \mathbb{J}$). We say that an object is "covered" in a given solution if the coverage requirement for it is satisfied in the solution. The *mutual exclusivity* constraints pertain to pairs of objects in $\mathbb{O}$ that must not be assigned to a same given $\mathbb{F}_j$ ($j \in \mathbb{J}$). Similarly, the *joint inclusion* constraints pertain to pairs of objects in $\mathbb{O}$ that must be assigned together (i.e., either they are both assigned to a same given $\mathbb{F}_j$ ($j \in \mathbb{J}$), or neither of them is assigned to any $\mathbb{F}_j$ ($j \in \mathbb{J}$)).

Example 4.3. *In reference to Example* 4.1:

- *If objects "1" and "2" were subject to a mutual exclusivity constraint, then the only feasible subsets would be* $\mathbb{F}_2$, $\mathbb{F}_3$, *and* $\mathbb{F}_4$.
- *If objects "1" and "2" were subject to a joint inclusion constraint, then the only feasible subsets would be* $\mathbb{F}_1$, *and* $\mathbb{F}_5$. $\qquad\square$

As discussed earlier in this section, a "natural" condition that applies to contexts in which the set of *crew subsets* is nonempty is that an object in $\mathbb{O}$ cannot be covered more than once (since that would imply the participation of the object to several crews simultaneously). Hence, in our framework each object in $\mathbb{O}$ will be required to be covered exactly once whenever the set of *crew subsets* is non-empty.

We distinguish between two broad classes of COPs which we refer to as "simple-cover" COPs (SCCOPs), and "weighted-cover" COPs (WCCOPs), respectively. In SCCOPs, the $\mathbb{F}_j s$ ($j \in \mathbb{J}$) have "weights" that are either 0 or 1 in the coverage requirement constraints. In WCCOPs, those weights are subject to non-negativity restrictions only. The basic (and simplest) of the SCCOPs is the Bipartite Matching Problem (BMP; see Burkhard *et al.* (2009)). However, the SCCOP class also includes many of the well-known hard (NP-Complete) COPs (see Karp, 1972; Garey and Johnson, 1979; Nemhauser and Wolsey, 1988, among others). The prototypical WCCOP is the 0/1 Knapsack Problem (see Karp, 1972, or Garey and Johnson, 1979, Nemhauser and Wolsey, 1988, among others).

The focus in this book is the SCCOP class. An illustration of this class is provided in Table 4.1 (where $\mathbb{E}$ and $\mathbb{I}$ denote the sets of object pairs that are subject to *mutual exclusivity*, and *joint inclusion* constraints, respectively) using the Bipartite Matching Problem (BMP), Traveling Salesman Problem (TSP), Vertex Coloring Problem (VCP), Set Partioning Problem (SPP), and Multiple Traveling Salesman Problem (mTSP; see Diaby (2010a)), respectively. In the BMP (without loss of generality), $\mathbb{O}$ consists of objects to be assigned to "positions"; $\mathfrak{s} = 1$; $\mathbb{J} = \{1\}$; $\mathbb{F}_1 = \mathbb{O}$; $\mathfrak{f}_1 = 0$ (i.e., there is no fixed cost); the "sequence-dependent" cost components are the object-position assignment costs; $\mathbb{E} = \varnothing$; $\mathbb{I} = \mathbb{O}^2$. In the TSP, $\mathbb{O}$ consists of cities to be visited; $\mathfrak{s} = 1$; $\mathbb{J} = \{1\}$; $\mathbb{F}_1 = \mathbb{O}\backslash\{1\}$ (assuming, without loss of generality, that city "1" has been designated as the beginning and ending point of the travels); $\mathfrak{f}_1 = 0$ (i.e., there is no fixed cost); the "sequence-dependent" costs are the inter-city travel costs/distances; $\mathbb{E} = \varnothing$; $\mathbb{I} = (\mathbb{O}\backslash\{1\})^2$. In the VCP (in this framework), $\mathbb{O}$ consists of nodes of a graph (the "VCP graph") to be "colored"; $\mathfrak{s}$ is equal to the number of available "colors"; $\mathbb{J} = \{1,\ldots,\mathfrak{s}\}$; $\forall j \in \mathbb{J}$, $\mathbb{F}_j = \mathbb{O}$; $\forall j \in \mathbb{J}$, $\mathfrak{f}_j = 1$; $\forall j \in \mathbb{J}$, $\forall(i,k) \in \mathbb{F}_j^2$, $\mathfrak{d}_{jik} = 0$ (i.e., there are no sequence-dependent costs); $\mathbb{E} = \{(i,j) \in \mathbb{O}^2 : i$ and j are adjacent in the VCP graph$\}$; $\mathbb{I} = \varnothing$. In the SPP, $\mathbb{O}$ consists of objects to be assigned to "object crews"; $\mathfrak{s}$ is equal to the number of given crews; $\mathbb{J} = \{1,\ldots,\mathfrak{s}\}$; the $\mathbb{F}_j's$ and the $\mathfrak{f}_j's$ are explicitly stated; $\forall j \in \mathbb{J}$, $\forall(i,k) \in \mathbb{F}_j^2$, $\mathfrak{d}_{jik} = 0$ (i.e., there are no sequence-dependent costs); $\mathbb{E} = \varnothing$; $\mathbb{I} = \varnothing$. In the multi-depot, fixed-destination mTSP, $\mathbb{O}$ consists of the cities to be visited; $\mathfrak{s}$ is the number of salesmen; $\mathbb{J} = \{1,\ldots,\mathfrak{s}\}$; $\forall j \in \mathbb{J}$, $\mathbb{F}_j = \mathbb{O}$; $\forall j \in \mathbb{J}$, $\mathfrak{f}_j$ is equal to the fixed cost associated with the "activation" of Salesman j; $\forall j \in \mathbb{J}$, $\forall(i,k) \in (\mathbb{F}_j \cup \{\mathbf{d}\})^2 : i \neq k$, $\mathfrak{d}_{jik}$ is the cost incurred when Salesman j travels from city i to city j (where $\mathbf{d} := (|\mathbb{O}|+1)$ is the index for a "dummy"/fictitious city representing depots/bases); $\mathbb{E} = \varnothing$; $\mathbb{I} = \varnothing$.

It is easy to verify that many of the other well-known hard COPs (see Karp (1972), or Garey and Johnson (1979), among others) can be readily "cast" into our proposed generalized framework as SCCOPs, using the approach illustrated in Table 4.1.

Table 4.1. Illustration of the generic description of SCCOPs.

SCCOP	$\mathbb{O}$	$\mathfrak{s}$	$\mathbb{J}$	$\mathbb{F}'_j s$	$\exists? f_j > 0$	$\exists? \partial_{jik} > 0$	$\mathbb{E}$	$\mathbb{I}$
BMP	Objects to assign to "positions"	1	$\{1\}$	$\mathbb{F}_1 = \mathbb{O}$	No	Yes (special-structured)	$\varnothing$	$\mathbb{O}^2$
TSP	Cities to "visit" in order	1	$\{1\}$	$\mathbb{F}_1 = \mathbb{O}\backslash\{1\}$	No	Yes	$\varnothing$	$(\mathbb{O}\backslash\{1\})^2$
VCP	Nodes to "colors"	# of "colors"	$\{1,\dots,\mathfrak{s}\}$	$\forall j \in \mathbb{J},$ $\mathbb{F}_j = \mathbb{O}$	Yes ($\forall j \in \mathbb{J}$, $f_j = 1$)	No	adjacent nodes pairs	$\varnothing$
SPP	Objects to assign to "crews"	# of crews	$\{1,\dots,\mathfrak{s}\}$	Explicit	Yes	No	$\varnothing$	$\varnothing$
mTSP	Subsets of cities to "visit" in order each	# of salesmen	$\{1,\dots,\mathfrak{s}\}$	$\forall j \in \mathbb{J},$ $\mathbb{F}_j = \mathbb{O}$	Yes	Yes	$\varnothing$	$\varnothing$

3. Generic Bipartite Network Flow-Based Model of SCCOP Solutions

We will now discuss a generic BNF based representation of SCCOP solutions which serves as a unifying reference in the generalization of our path representations of COP solutions exemplified in Section 3 for the TSP. Cells of the tableau underlying the proposed BNF-based formulation correspond to nodes of the flow graph for a given COP, whereas the specific side-constraints of the formulation for the given COP define the arcs of the flow graph of the COP.

Notation 4.1.

(1) $\mathfrak{u}$: Number of objects;

(2) $\mathbb{O} := \{1, 2, \ldots, \mathfrak{u}\}$ (Set of objects);

(3) $\mathfrak{s}$: Number of subsets of $\mathbb{O}$ under consideration;

(4) $\mathbb{J} := \{1, 2, \ldots, \mathfrak{s}\}$ (Index set for the subsets under consideration);

(5) $\overline{\mathbb{J}}$: Index set of the *crew subsets* of $\mathbb{O}$ ($\overline{\mathbb{J}} \subseteq \mathbb{J}$);

(6) $\mathbb{F} := \{\mathbb{F}_1, \ldots, \mathbb{F}_\mathfrak{s}\}$ (Set of subsets under consideration);

(7) $\forall j \in \mathbb{J}, \forall i \in \mathbb{O}$,

$$a_{ij} = \begin{cases} 1 & \text{if object } i \text{ can be assigned to subset } \mathbb{F}_j; \\ 0 & \text{otherwise;} \end{cases}$$

(Binary indicator for the memberships of the subsets under consideration);

(8) $\forall j \in \mathbb{J}, \mathfrak{m}_j := |\mathbb{F}_j|$ (Number of objects that can be assigned to $\mathbb{F}_j$);

(9) $\forall j \in \mathbb{J}, \mathfrak{T}_j := \{1, 2, \ldots, \mathfrak{m}_j\}$ ("Times" of assignments of objects for $\mathbb{F}_j$);

(10) $\mathfrak{n} := \sum_{j \in J} \mathfrak{m}_j$ (Number of non-zero entries of the SCCOP input matrix);

(11) $\forall j \in \mathbb{J}, \mathfrak{f}_j$: Fixed cost incurred if $\mathbb{F}_j$ is chosen;

(12) $\forall j \in \mathbb{J}, \forall (i, k) \in \mathbb{F}_j^2, \mathfrak{v}_{j,i,k}$: Cost incurred if $\mathbb{F}_j$ is chosen, with object k assigned immediately after object i.

Assumption 4.1. *We assume without loss of generality that:*

(1) *The set of objects has been augmented with a fictitious ("dummy") object, indexed as* $\mathbf{d} := \mathbf{u} + 1$;
(2) *The fictitious object can be assigned to every* $\mathbb{F}_j$, $j \in \mathbb{J}$ *(i.e.,* $\forall j \in \mathbb{J}$, $a_{\mathbf{d}j} = 1$*).*

Notation 4.2. $\forall j \in \mathbb{J}$, $\forall t \in \mathfrak{T}_j$, $\forall i \in \mathbb{O} \cup \{\mathbf{d}\}$, $v_{j,t,i} : 0/1$ *binary variable that is equal to 1 iff* i *is the tth object assigned to* $\mathbb{F}_j$.

Definition 4.1. *Let* $P_1 := \{v \in \mathbb{R}^{n^2} : v \text{ satisfies } (4.1)\text{--}(4.6)\}$, *where* (4.1)–(4.6) *are specified as follows:*

$$\sum_{j \in \mathbb{J}} \sum_{t \in \mathfrak{T}_j} a_{ij} v_{j,t,i} = |\geq 1; \quad i \in \mathbb{O}, \tag{4.1}$$

$$\sum_{j \in \mathbb{J}} \sum_{t \in \mathfrak{T}_j} a_{ij} v_{j,t,\mathbf{d}} \leq \sum_{j \in \mathbb{J}} |\mathbb{F}_j| - |\mathbb{O}|, \tag{4.2}$$

$$\sum_{i \in (\mathbb{O} \cup \{\mathbf{d}\})} a_{ij} v_{j,t,i} = 1; \quad j \in \mathbb{J}, \ t \in \mathfrak{T}_j, \tag{4.3}$$

$$\sum_{i \in \mathbb{O}} a_{ij} v_{j,t,i} - \sum_{i \in \mathbb{O}} a_{ij} v_{j,t-1,i} \leq 0, \quad j \in \mathbb{J} \backslash \bar{\mathbb{J}}, \ t \in \mathfrak{T}_j \backslash \{1\}, \tag{4.4}$$

$$\sum_{i \in \mathbb{O}} a_{ij} v_{j,t,i} - \sum_{i \in \mathbb{O}} a_{ij} v_{j,1,i} = 0, \quad j \in \bar{\mathbb{J}}, \ t \in \mathfrak{T}_j \backslash \{1\} \tag{4.5}$$

$$v_{j,t,i} \in \{0,1\}; \quad j \in \mathbb{J}, \ t \in \mathfrak{T}_j, \ i \in (\mathbb{O} \cup \{\mathbf{d}\}). \tag{4.6}$$

We refer to $\mathrm{Conv}(P_1)$ as the "BNF-based polytope".

Constraints (4.1) require that each object in $\mathbb{O}$ be assigned to exactly ("=") or at least one ("$\geq$") of the $\mathbb{F}_j$'s. Constraints (4.3) ensure that each *assignment possibility* for a given $\mathbb{F}_j$ ($j \in \mathbb{J}$) is used at most once. Constraints (4.4) require that all the assignments to a given $\mathbb{F}_j$ ($j \in \mathbb{J}$) are made at consecutive "times". Constraints (4.5) enforce the additional assignment requirement for *crew subsets*. Note that for any $j \in \mathbb{J}$, constraints (4.4) would be implied by constraints (4.5). Hence, those constraints need to be stated for the *non-crew subsets* only, as shown above.

		F_1					F_2					$\cdots$	F_s					
		1	2	$\cdots$	$m_1{-}1$	m_1	1	2	$\cdots$	$m_2{-}1$	m_2	$\cdots$	1	2	$\cdots$	$m_5{-}1$	m_5	"Demand"
O	1			$\cdots$					$\cdots$			$\cdots$			$\cdots$			1
	2			$\cdots$					$\cdots$			$\cdots$			$\cdots$			1
	$\vdots$	$\vdots$	$\vdots$	$\vdots$	$\vdots$	$\vdots$	$\vdots$	$\vdots$	$\vdots$	$\vdots$	$\vdots$	$\vdots$	$\vdots$	$\vdots$	$\vdots$	$\vdots$	$\vdots$	$\vdots$
	$o{-}1$			$\cdots$					$\cdots$			$\cdots$			$\cdots$			1
	o			$\cdots$					$\cdots$			$\cdots$			$\cdots$			1
{d}	$o{+}1$			$\cdots$					$\cdots$			$\cdots$			$\cdots$			r
"Supply"		1	1	$\cdots$	1	1	1	1	$\cdots$	1	1	$\cdots$	1	1	$\cdots$	1	1	

$$r := \sum_{j \in J} |F_j| - |O|$$

Figure 4.1. Tableau form for the generic BNF model.

The generic form of the BNF tableau underlying P_1 is illustrated in Figure 4.1. The specific cases of the TSP, SPP, VCP, and mTSP are illustrated in Examples 4.4–4.7 respectively.

Example 4.4 (Traveling Salesman Problem (TSP)). *For the Traveling Salesman Problem with 10 cities:*

- *The "objects" are the cities "2" through "10" (i.e., $\mathbb{O} = \{2, 3, 4, 5, 6, 7, 8, 9, 10\}$) (assuming city "1" is the starting and ending city);*
- $s = 1$;
- $\mathbb{F}_1 = \mathbb{O} = \{2, 3, 4, 5, 6, 7, 8, 9, 10\}$;
- $\mathbb{F}_1$ *can be treated either as a "free subset" or a "crew subset";*
- *The "dummy" city is not needed (since the right-hand-side of constraint (4.2) for it is zero);*
- *The BNF tableau underlying P_1 is shown in Table 4.2.*

Example 4.5 (Set Partitioning Problem (SPP)). *For the set partitioning problem with*

- $\mathbb{O} = \{1, 2, 3, 4, 5, 6\}$;
- $d = 7$;
- $s = 5$;
- $\mathbb{F}_1 = \{1, 3, 5\}$,
  $\mathbb{F}_2 = \{2, 6\}$,
  $\mathbb{F}_3 = \{2\}$,

Table 4.2. Generic BNF tableau for the TSP.

| | | $\mathbb{F}_1$ | | | | | | | | | |
		1	2	3	4	5	6	7	8	9	"Demand"
	2	1	1	1	1	1	1	1	1	1	1
	3	1	1	1	1	1	1	1	1	1	1
	4	1	1	1	1	1	1	1	1	1	1
	5	1	1	1	1	1	1	1	1	1	1
$\mathbb{O}$ (cities)	6	1	1	1	1	1	1	1	1	1	1
	7	1	1	1	1	1	1	1	1	1	1
	8	1	1	1	1	1	1	1	1	1	1
	9	1	1	1	1	1	1	1	1	1	1
	10	1	1	1	1	1	1	1	1	1	1
"Supply"		1	1	1	1	1	1	1	1	1	—

Table 4.3. Generic BNF tableau for the SPP.

| | | $\mathbb{F}_1$ | | | $\mathbb{F}_2$ | | $\mathbb{F}_3$ | $\mathbb{F}_4$ | $\mathbb{F}_5$ | | | $\mathbb{F}_6$ | | $\mathbb{F}_7$ | | | $\mathbb{F}_8$ | | "Demand" |
		1	2	3	1	2	1	1	1	2	3	1	2	1	2	3	1	1	
	1	1	1	1					1	1	1	1	1	1	1	1			1
	2				1	1	1		1	1	1						1	1	1
$\mathbb{O}$	3	1	1	1								1	1						1
	4							1						1	1	1			1
	5	1	1	1										1	1	1	1	1	1
	6						1	1	1	1	1								1
$\{\mathbf{d}\}$	7	1	1	1	1	1	1	1	1	1	1	1	1	1	1	1	1	1	11 (= 17 − 6)
"Supply"		1	1	1	1	1	1	1	1	1	1	1	1	1	1	1	1	1	—

$\mathbb{F}_4 = \{4\}$,
$\mathbb{F}_5 = \{1, 2, 6\}$,
$\mathbb{F}_6 = \{1, 3\}$,
$\mathbb{F}_7 = \{1, 4, 5\}$,
$\mathbb{F}_8 = \{2, 5\}$;

- Each $\mathbb{F}_j$ $(j \in \mathbb{J} = \{1, 2, 3, 4, 5, 6, 7, 8\})$ *is a* crew subset;
- *The Assignment tableau underlying* P_1 *is shown in Table* 4.3.

Example 4.6 (Vertex Coloring Problem (VCP)). *For the vertex coloring problem with three colors and the graph below,*

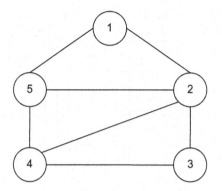

- *The "objects" are the vertices (i.e., $\mathbb{O} = \{1, 2, 3, 4, 5\}$);*
- *The "subsets" are the colors. Hence, $\mathfrak{s} = 3$;*
- $\mathbb{F}_j = \{1, 2, 3, 4, 5\}$, $j = 1, 2, 3$;
- *Each $\mathbb{F}_j$ $(j = 1, 2, 3)$ is a free subset;*
- $\mathbf{d} = 6$;
- *The BNF tableau underlying P_1 assuming a given color is assigned in order with the lower indexed object "receiving" the color first is shown in Table 4.4.*

Table 4.4. Generic BNF tableau for the VCP.

		$\mathbb{F}_1(c_1)$					$\mathbb{F}_2(c_2)$					$\mathbb{F}_3(c_3)$					*"Demand"*
		1	2	3	4	5	1	2	3	4	5	1	2	3	4	5	
$\mathbb{O}$	1	1					1					1					1
	2	1	1				1	1				1	1				1
(vertices)	3	1	1	1			1	1	1			1	1	1			1
	4	1	1	1	1		1	1	1	1		1	1	1	1		1
	5	1	1	1	1	1	1	1	1	1	1	1	1	1	1	1	1
$\{\mathbf{d}\}$	6	1	1	1	1	1	1	1	1	1	1	1	1	1	1	1	$10(= 15 - 5)$
"Supply"		1	1	1	1	1	1	1	1	1	1	1	1	1	1	1	–

Example 4.7 (Multiple Traveling Salesman Problem (MTSP)). *For a multi-depot, fixed-destination, multiple TSP with two salesmen and five cities:*

- *The "objects" are the cities (i.e., $\mathbb{O} = \{1, 2, 3, 4, 5\}$);*
- $\mathbf{d} = 6$;
- *The "subsets" are the salesmen. Hence, $\mathfrak{s} = 2$;*
- $\mathbb{F}_j = \{1, 2, 3, 4, 5\}$, $j = 1, 2$;

Table 4.5. Generic BNF tableau for the mTSP.

		$\mathbb{F}_1$ (*Salesman 1*)					$\mathbb{F}_2$ (*Salesman 2*)					*"Demand"*
		1	2	3	4	5	1	2	3	4	5	
	1	1	1	1	1	1	1	1	1	1	1	1
	2	1	1	1	1	1	1	1	1	1	1	1
$\mathbb{O}$ (*cities*)	3	1	1	1	1	1	1	1	1	1	1	1
	4	1	1	1	1	1	1	1	1	1	1	1
	5	1	1	1	1	1	1	1	1	1	1	1
$\{\mathbf{d}\}$	6	1	1	1	1	1	1	1	1	1	1	5 (= 10 − 5)
"Supply"		1	1	1	1	1	1	1	1	1	1	−

- Each $\mathbb{F}_j$ ($j = 1, 2$) *is a free subset;*
- *The Assignment tableau underlying* P_1 *is shown in Table* 4.5.

Theorem 4.1.

(1) *Every point of P_1 (i.e., every extreme point of the BNF-based polytope) corresponds to a solution of the generic SCCOP;*
(2) *Every solution of the generic SCCOP corresponds to a point of P_1;*
(3) *The mapping of points of P_1 to the solutions of the generic SCCOP is many-to-one if there are no sequence-dependent costs and no particular ordering is imposed in then modeling;*
(4) *The mapping of points of P_1 to the solutions of the generic SCCOP is one-to-one if there are sequence-dependent costs.*

Proof. It is trivial to construct a point of P_1 from any given SCCOP solution and vice versa. Statements (1) and (2) of the theorem follow from this directly. Statement (3) and (4) follow from the fact that the order in which the non-fictitious objects are assigned can be changed without changing the corresponding SCCOP solution if there are no sequence-dependent costs. □

4. Generic Flow Graphs (GFG)

In order to state the linear program for our proposed generic SCCOP, we express the generic SCCOP as a flow problem, as briefly discussed in Chapter 1 and fully developed in Section 3. The structure of the flow graph we use for this is illustrated in Figure 4.2. We refer to this

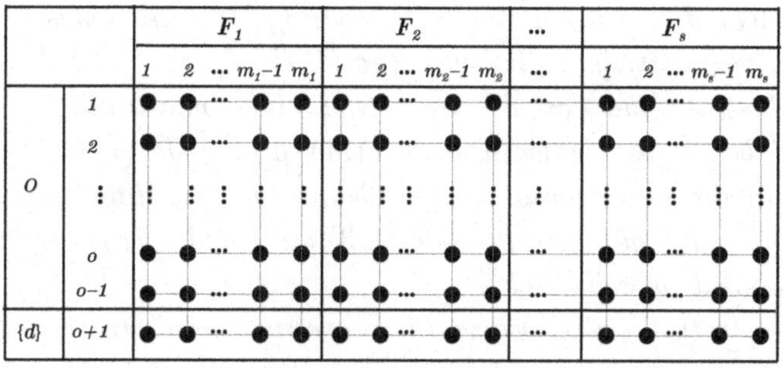

		F_1				F_2				•••	F_s			
		1	2	••• m_1-1	m_1	1	2	••• m_2-1	m_2	•••	1	2	••• m_s-1	m_s
O	1	●	●	••• ●	●	●	●	••• ●	●	•••	●	●	••• ●	●
	2	●	●	••• ●	●	●	●	••• ●	●	•••	●	●	••• ●	●
	⋮	⋮	⋮ ⋮ ⋮	⋮	⋮	⋮	⋮ ⋮	⋮	⋮	⋮	⋮	⋮ ⋮	⋮	⋮
	o	●	●	••• ●	●	●	●	••• ●	●	•••	●	●	••• ●	●
	$o-1$	●	●	••• ●	●	●	●	••• ●	●	•••	●	●	••• ●	●
$\{d\}$	$o+1$	●	●	••• ●	●	●	●	••• ●	●	•••	●	●	••• ●	●

- The specific nodes to be included depend on the memberships of the F_j's;
- The specific arcs to be included depend on the applicable sets I and E.

Figure 4.2. Illustration of the structure of the GFG.

graph as the "GFG". The nodes of the GFG correspond to the "cells" of the tableau of the BNF-based formulation defined by (4.1)–(4.6), and illustrated in Examples 4.4–4.7.

The arcs of the GFG link nodes at consecutive *stages* of the graph. However, the specific arcs to be included in the graph depend on the specific forms of the side-constraints (4.4) and (4.5) for the SCCOP at hand. The general rule for enforcing the sequencing side-constraints (4.4) is to exclude arcs originating at fictitious nodes and ending at non-fictitious nodes, except for the last *stage* for the subset $\mathbb{F}_j$ ($j \in \mathbb{J}$) at hand. The general rule for enforcing the "crew" restriction side-constraints (4.5) is to exclude arcs originating at non-fictitious nodes and ending at fictitious nodes, except for the last *stage* for the subset $\mathbb{F}_j$ ($j \in \mathbb{J}$) concerned. In addition to these, the mutual exclusivity constraints are handled by not including arcs linking non-fictitious nodes of the graph if the corresponding objects are subject to mutual exclusivity constraints.

Example 4.8. *Consider the SPP described in Example* 4.5. *The rules described above, can be illustrated for this problem as follows:*

- *To illustrate the sequencing side-constraints* (4.4), *there would be no arc in the GFG between the node corresponding to cell* $(7,1)$ *("node* $(7,1)$*") and any of the nodes* $(1,2)$, $(3,2)$, *and* $(5,2)$, *for example. On the other hand, Stage* 3 *being the last stage for* $\mathbb{F}_1$, *there would be arcs going from node* $(7,3)$ *into nodes* $(2,4)$ *and* $(6,4)$, *respectively, included in the GFG;*

- *To illustrate the "crew" restriction side-constraints* (4.5), *there would be no arc between node* $(1,1)$ *and node* $(7,2)$, *for example. On the other hand, Stage 3 being the last stage for* $\mathbb{F}_1$, *there would be arcs going from node* $(1,3)$ *into node* $(7,4)$, *which would be included in the GFG;*

- *To illustrate the case where there are no "crew" restrictions (i.e., when there are "free subsets" as in the case of the Set Covering Problem, for example), there would be an arc between node* $(1,1)$ *and node* $(7,2)$, *for example, which would be included in the GFG.*

5. Overall LP Models for SCCOPs

After the GFG for a specific SCCOP at hand has been formalized, as illustrated in Notation 1.1 for the TSP, the specific constraints of our proposed "complex flow" model, as stated in Chapter 2, can then be written for the given SCCOP. Clearly, by the developments in Chapter 3, the LP relaxation of this model is integral. Hence, finally, in order to complete the formulation of the SCCOP at hand as a linear program, alternative linear cost functions in the (y- and/or z-) modeling variables can be developed following the approach we have proposed in Section 5 of Chapter 3. These ideas will be illustrated more specifically in Chapter 7 of this book.

Chapter 5

Non-Symmetry of the Basic (TSP) Model

1. Introduction

Yannakakis (1991) defines the notion of a *symmetric polytope* relative to the symmetric group on the set of cities (Ω), as follows:

> "We say that a polytope $P(x, y)$ over variables $x = (x_{ij})$ and new variables y is *symmetric* if every permutation π of the nodes can be also extended to the new variables y so that P remains invariant. A LP (set of linear constraints) is called symmetric if its feasible space is." (Yannakakis (1991, p. 447)).
>
> "...Informally, "symmetry" means that the nodes of the complete graph are treated the same way..." (Yannakakis (1991, p. 442)).

In this chapter, we will show that Q_L is not a symmetric model, and that it cannot be extended into a symmetric model using the *traditional x-variables*.

2. Non-Symmetry of the Basic Model

Theorem 5.1. Q_L *is not a symmetric polytope.*

Proof. The theorem follows directly from the observation that every relabeling of the nodes that changes the label of node "1" also changes the variables and constraints of Q_L in such a way as for them to describe a polytope that is distinct from Q_L.

More specifically, suppose the TSPFG is expanded to include a level (but not a stage) for (the TSP) node 1, so that the y- and

z-variables involving node 1 (which are implicitly set equal to 0 in the description of Q_L) are allowed to be explicitly expressed. Let $\binom{\bar{y}}{\bar{z}}$ denote the resulting "expanded" vector of y- and z-variables. To simplify the presentation, for $i \in \Omega$, denote by $\mathbf{u}_i$ the vector of components of $\binom{\bar{y}}{\bar{z}}$ that involve node i, and by $\mathbf{v}_i$ the vector of components that do not involve node i. Then, clearly, the constraints of Q_L can be rearranged based on a given node $k \in \Omega \backslash \{1\}$, and stated accordingly in the general ("partitioned-matrix") form (see Cullen (1972, pp. 31–37), among others):

$$\begin{bmatrix} \mathbf{A}_k & \mathbf{B}_k & \mathbf{0} \\ \mathbf{0} & \mathbf{0} & \mathbf{I} \end{bmatrix} \begin{bmatrix} \mathbf{u}_k \\ \mathbf{v}_k \\ \mathbf{u}_1 \end{bmatrix} = \begin{bmatrix} \mathbf{b} \\ \mathbf{0} \end{bmatrix}, \tag{5.1}$$

where $\mathbf{A}_k$ is the matrix of "technical" coefficients for the variables involving node k in the description of Q_L (over the expanded TSPFG); $\mathbf{B}_k$ is the matrix of "technical" coefficients for the variables not involving node k in the description of Q_L; $\mathbf{I}$ is an identity matrix of comfortable dimension; and $\mathbf{b}$ is the column vector of "right-hand-side" values for Q_L.

Clearly, $\forall (i,j) \in \Omega^2 : i \neq j$, $\mathbf{u}_i$ and $\mathbf{u}_j$ can be arranged so that $\mathbf{A}_i = \mathbf{A}_j = \tilde{\mathbf{A}}_{ij}$. Further, $\forall (i,j) \in \Omega^2 : i \neq j$, let $\tilde{\mathbf{v}}_{ij}$ denote the vector of components of $\binom{\bar{y}}{\bar{z}}$ which respectively involve neither nodes i nor j. Then, because the constraints

$$\mathbf{I} \cdot \mathbf{u}_1 = \mathbf{u}_1 = \mathbf{0} \tag{5.2}$$

are *valid* for Q_L, (5.1) can be rewritten as:

$$\begin{bmatrix} \tilde{\mathbf{A}}_{1,k} & \tilde{\mathbf{B}}_{1,k} & \mathbf{0} \\ \mathbf{0} & \mathbf{0} & \mathbf{I} \end{bmatrix} \begin{bmatrix} \mathbf{u}_k \\ \tilde{\mathbf{v}}_{1,k} \\ \mathbf{u}_1 \end{bmatrix} = \begin{bmatrix} \mathbf{b} \\ \mathbf{0} \end{bmatrix}, \tag{5.3}$$

where $\tilde{\mathbf{B}}_{1,k}$ is obtained from $\mathbf{B}_k$ by deleting the columns of $\mathbf{B}_k$ that pertain to components of $\mathbf{v}_k$ involving node 1.

Clearly, we have that:

$$\forall (i,j) \in \Omega^2 : i \neq j, \quad \tilde{\mathbf{A}}_{i,j} = \tilde{\mathbf{A}}_{j,i}; \quad \tilde{\mathbf{B}}_{i,j} = \tilde{\mathbf{B}}_{j,i}; \quad \text{and} \quad \tilde{\mathbf{v}}_{i,j} = \tilde{\mathbf{v}}_{j,i}. \tag{5.4}$$

Now, consider a relabeling, $\mathcal{R}$, of the TSP nodes in which the node/level labels "1" and "k" are swapped (with all the other labels maintained). Let the resulting polytope be $\widehat{Q}_L = \mathcal{R}(Q_L)$. Then, using (5.4), and re-arranging $\widehat{Q}_L$ according to node/level 1, $\widehat{Q}_L$ can be stated as:

$$\begin{bmatrix} \widetilde{\mathbf{A}}_{1,k} & \widetilde{\mathbf{B}}_{1,k} & \mathbf{0} \\ \mathbf{0} & \mathbf{0} & \mathbf{I} \end{bmatrix} \begin{bmatrix} \mathbf{u}_1 \\ \widetilde{\mathbf{v}}_{1,k} \\ \mathbf{u}_k \end{bmatrix} = \begin{bmatrix} \mathbf{b} \\ \mathbf{0} \end{bmatrix}. \tag{5.5}$$

Hence, the relabeling essentially swaps the "technical" coefficients of corresponding components of $\mathbf{u}_k$ and $\mathbf{u}_1$, respectively. Hence, clearly, $\widehat{Q}_L$ and Q_L are congruent polytopes. Note however, that because of the *visit requirements* constraints (2.8), constraints (5.2) are infeasible for $\widehat{Q}_L$, and constraints

$$\mathbf{I} \cdot \mathbf{u}_k = \mathbf{u}_k = \mathbf{0} \tag{5.6}$$

are infeasible for Q_L. Hence, $\widehat{Q}_L$ and Q_L do not coincide. Hence, $\widehat{Q}_L$ and Q_L are distinct, and respectively, non-symmetric. The theorem follows from this directly. $\qquad\square$

The concepts used in the proof of Theorem 5.1 are illustrated in Figures 5.1 and 5.2. Figure 5.1 illustrates the case of a "swap" of

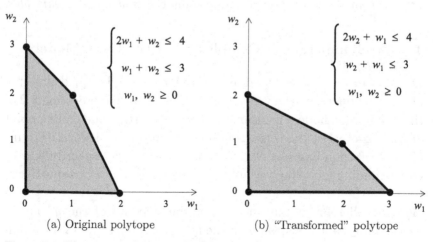

(a) Original polytope   (b) "Transformed" polytope

Figure 5.1. Illustration of the proof of Theorem 5.1 for a non-symmetric polytope.

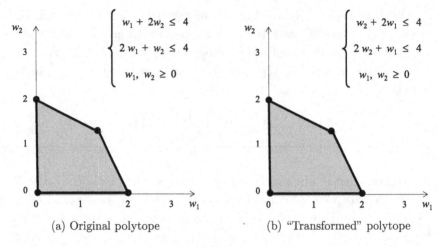

(a) Original polytope   (b) "Transformed" polytope

Figure 5.2. Illustration of the proof of Theorem 5.1 for a symmetric polytope.

constraint matrix coefficients between variables when the polytope described is non-symmetric. As can be seen from this figure, the "original" and "transformed" polytopes are congruent, but do not coincide. In Figure 5.2 (where the "original" polytope is symmetric (around the 45-degree line, specifically)), the "original" and "transformed" polytopes are congruent, and also coincide.

We will show in the next set of results, that Q_L cannot be *extended* into a symmetric polytope using the *traditional x-variables*.

3. Non-Symmetry of "Complexes" of the Basic Model

Definition 5.1 ("FSC-based model"). For $\alpha \in \Omega$, let $\widetilde{Q}_L^\alpha$ denote the polytope induced by the version of the model stated in Section 5.2 of this book when node α is assumed to be the starting and ending point of the travels. Let the corresponding flow graph be defined in terms of the corresponding sets $\widetilde{N}_r^\alpha$, $\widetilde{F}_r^\alpha$, and $\widetilde{B}_r^\alpha$ ($r \in S$), respectively, with corresponding modeling variables denoted as $\widetilde{y}^\alpha$ and $\widetilde{z}^\alpha$, respectively.

(1) We will refer to any model that includes all of the $\widetilde{y}^\alpha$ and $\widetilde{z}$ variables and constraints of the $\widetilde{Q}_L^\alpha$'s ($\alpha \in \Omega$), as a "Finite Complex (FC)" model (since, the structure of $\{\widetilde{Q}_L^\alpha, \ \alpha = 1, \ldots, n\}$ is

somehow similar to that of a finite simplicial complex (see Panik (1993, pp. 239–251), among others);

(2) We will refer to the polytope induced by an *FC model* as an "FC polytope".

Remark 5.1. *The model consisting of the complex* $\{\widetilde{Q}_L^\alpha, \alpha = 1, \ldots, n\}$ *only is an FC model.*

Theorem 5.2. *A symmetric FC polytope cannot be an equivalent of* Q_L.

Proof. Let TSP tours be represented using $(n + 1)$-tuples $(i_1, i_2, \ldots, i_n, i_1)$, where: $i_p \in \Omega$ for $p = 1, \ldots, n$; and $i_p \neq i_q$ for $(p, q) \in \Omega^2 : p \neq q$. Consider any given *FC model*, say "Model FC". Assume that this model is symmetric. Then, we have the following:

(a) By the symmetry of *Model FC*, each $(n + 1)$-tuple representation, $(i_k, i_{k+1}, \ldots, i_{k+n})$ $(k = 1, \ldots, n; i_{n+k} := i_k)$, of a given TSP tour, $(i_1, i_2, \ldots, i_n, i_1)$, must correspond to a feasible solution of *Model FC*. Hence, every TSP tour must have (feasible) fractional-solution representations in *Model FC* (since any convex combination of feasible solutions of *Model FSC* must be a feasible solution to *Model FC*). However, clearly, a TSP tour does not have any feasible fractional representation in Q_L.

(b) Because of its symmetry, *Model FC* must include constraints that "force" a consistency in a given feasible solution to the model, among the $(\widetilde{y}^\alpha, \widetilde{z}^\alpha)$'s $(\alpha \in \Omega)$ in terms of the TSP tours they respectively represent (since, absent any such constraints, it would be possible for a given solution to the model to represent distinct tours (at the same time)). Because of these "complicating" "side-constraints" in particular, the overall (FC) polytope cannot be an equivalent of any single one of the $\widetilde{Q}^\alpha$'s $(\alpha \in \Omega)$.

The theorem follows directly from (a) and (b) above. $\qquad\square$

The proof of Theorem 5.2 is illustrated in Example 5.1.

Example 5.1. *Theorem 5.2 can be illustrated using the case of $n = 6$ as follows:*

(a) *Consider the TSP tour $(3-6-5-1-4-2-3)$. This tour is represented as $(1-4-2-3-6-5-1)$ in $\widetilde{Q}_L^1$; as $(2-3-6-5-1-4-2)$ in $\widetilde{Q}_L^2$; as $(3-6-5-1-4-2-3)$ in $\widetilde{Q}_L^3$; etc. Let $(\widetilde{y}^1, \widetilde{z}^1)^*$, $(\widetilde{y}^2, \widetilde{z}^2)^*$, $(\widetilde{y}^3, \widetilde{z}^3)^*$, etc., be the corresponding characteristic vectors in a given (symmetric) FC model, "Model FC." Clearly, any convex combination of these $(\widetilde{y}^k, \widetilde{z}^k)^*$'s $(k = 1, \ldots, n)$, must be a feasible solution to Model FC. Hence, "tour $(3-6-5-1-4-2-3)$" has fractional-solution representations in Model FC, which it does clearly not have in Q_L.*

(b) *There must exist constraints in Model FC that enforce the consistency between the tours represented "within" a given solution of the model. If not, then, for example, a solution vector in which the $(\widetilde{y}^1, \widetilde{z}^1)^*$ components correspond to the tour, say $(1-4-2-3-6-5-1)$, while the $(\widetilde{y}^2, \widetilde{z}^2)^*$ components correspond to the tour, say $(2-1-6-5-3-4-2)$, would be feasible for Model FC.*

Theorem 5.3. *Q_L cannot be extended into a symmetric polytope using the traditional x-variables.*

Proof. First, observe (see the proof of Theorem 5.1) that the non-symmetry of Q_L is due (at least, in part) to the changes in the "descriptive" y- and z-variables that results from relabelings of the TSP nodes. Note also that an individual y- or z-variable cannot be expressed as a linear function of the *traditional x-variables*. Hence, the non-symmetry of Q_L cannot be changed using the *traditional x-variables*. The theorem follows directly from the combination of this and Theorem 5.2. $\square$

Chapter 6

Non-Applicability of *Extended Formulations* Theory

1. Introduction

There has been a renewed interest and excellent work in Extended Formulations (EFs) over the past 4 years (see Conforti *et al.* (2010), Vanderbeck and Wolsey (2010), Fiorini *et al.* (2011; 2012), and Kaibel (2011), for example). Despite the great importance of the EFs paradigm in the analysis of linear programming (LP) and integer programming (IP) models of combinatorial optimization problems (COPs), the clear definition of its scope of applicability has been a somewhat-overlooked issue. The purpose of this chapter is to make a contribution towards addressing this issue. Specifically, we will show that the notion of an EF can become *ill-defined* and degenerate (and thereby lose its meaningfulness) when it is being used to relate polytopes involved in alternate abstractions (see Borowski and Borwein (1991, p. 4) for a formal definition) of a given optimization problem. Because most of the papers on EFs focus on the TSP specifically, we will center our discussion on the TSP. However, the substance of the chapter is applicable to other NP-Complete problems.

Our aim is to bring attention to limits to the scope within which EFs Theory is applicable when attempting to derive bounds on the size of a LP model. The following remark is a recall of the fact that the TSP optimization problem can be modeled independently of/without any reference to the Standard TSP polytope, which is important for our purpose.

Remark 6.1.

(1) *Alternate abstractions (see Borowski and Borwein (1991, p. 4))*
of the TSP optimization problem which may or may not involve
the Standard TSP Polytope are possible. For example, in the
"standard" (i.e., Standard TSP Polytope-based) abstraction of
the TSP optimization problem, the cost function is trivial to
develop. The challenge is to come up with linear constraints so
that the extreme points of the induced polytope are TSP tours.
On the other hand, in an abstraction based on the Assignment
Polytop (AP; see Theorem 2.1), the representation of the TSP
tours is a straightforward matter (since the tours are abstracted
into assignment problem (bipartite matching) solutions). The
challenge is to find appropriate costs to apply to these thus-
abstracted TSP tours. Clearly, this challenge of coming up with
a cost function is not within the scope of EFs developments for
the Standard TSP Polytope, since it does not involve that poly-
tope. Examples of how this challenge can be met are described
in Diaby (2007) and in Chapters 2–3 of this book, respectively.
Also, clearly, from an overall perspective, one cannot reasonably
equate an "impossibility" of meeting the challenge in one of the
two abstractions (i.e., the "Standard TSP Polytope-based" and
"AP-based" abstractions) with an "impossibility" of meeting the
challenge in the other.

(2) *It follows directly from Theorem 2.1 that AP is a contradiction of*
any notion whereby it is "impossible" to abstract the TSP poly-
tope into a linear program of polynomial size, since AP clearly
has polynomial (linear) size and it is a well-established fact that
AP is integral (see Burkard et al. (2009)).

The plan of the chapter is as follows. First, in Section 2, we will
review the basic definitions and notation. Then, we will introduce
the notion of "polyhedron augmentation" in Section 3.1, and use it
(in Section 3.2) to develop our results on the condition about EFs
becoming *ill-defined*. In Section 4, we elaborate on our results and
demonstrate their applicability in the case of the Minimum Spanning
Tree Problem (MSTP).

2. Background Overview

2.1. *Basic definitions*

Definition 6.1 ("Traditional x-Variables"). We will generically refer to 2-indexed variables that have been used in traditional IP formulations of the TSP to represent inter-city travels as "traditional x-variables". In other words, $\forall (i,j) \in \{1,\ldots,n\}^2 : i \neq j$, we will refer to the 0/1 decision variables that are such that the (i,j)th entry of their vector is equal to "1" iff there is travel from city i to city j, as the "traditional x-variables", irrespective of the symbol used to denote them.

Definition 6.2 ("Standard EF Definition" (Yannakakis (1991); Conforti *et al.* (2010; 2013))). An EF for a polytope $X \subseteq \mathbb{R}^p$ is a polyhedron $U = \{(x,w) \in \mathbb{R}^{p+q} : Gx + Hw \leq g\}$ the projection, $\varphi_x(U)$ $:= \{x \in \mathbb{R}^p : (\exists w \in \mathbb{R}^q : (x,w) \in U)\}$, of which onto x-space is equal to X (where $G \in \mathbb{R}^{m \times p}$, $H \in \mathbb{R}^{m \times q}$, and $g \in \mathbb{R}^m$).

Definition 6.3 ("Alternate EF Definition #1" (Kaibel (2011); Fiorini *et al.* (2011; 2012))). A polyhedron $U = \{(x,w) \in \mathbb{R}^{p+q} : Gx + Hw \leq g\}$ is an EF of a polytope $X \subseteq \mathbb{R}^p$ if there exists a linear map $\pi : \mathbb{R}^{p+q} \longrightarrow \mathbb{R}^p$ such that X is the image of U under π (i.e., $X = \pi(U)$; where $G \in \mathbb{R}^{m \times p}$, $H \in \mathbb{R}^{m \times q}$, and $g \in \mathbb{R}^m$).

Definition 6.4 ("Alternate EF Definition #2" (Fiorini *et al.* (2012))). An EF of a polytope $X \subseteq \mathbb{R}^p$ is a linear system $U = \{(x,w) \in \mathbb{R}^{p+q} : Gx + Hw \leq g\}$ such that $x \in X$ if and only if there exists $w \in \mathbb{R}^q$ such that $(x,w) \in U$. (In other words, U is an EF of X if $(x \in X \Longleftrightarrow (\exists w \in \mathbb{R}^q : (x,w) \in U)))$ (where $G \in \mathbb{R}^{m \times p}$, $H \in \mathbb{R}^{m \times q}$, and $g \in \mathbb{R}^m$).

Remark 6.2. *The following observations are in order with respect to Definitions 6.2, 6.3, and 6.4:*

(1) *The statement of U in terms of inequality constraints only does not cause any loss of generality, since each equality constraint can be replaced by a pair of inequality constraints. (Yannakakis*

(1991, p. 442), *for example, just says that U is a set of linear constraints.*)

(2) *The statements "U is an EF of X" and "U expresses X" (see Yannakakis (1991)) are equivalent.*

(3) *The system of linear equations which specify π in Definition 6.3 must be valid constraints for X and U. Hence, X and U can be respectively extended by adding those constraints to them, when trying to relate X and U using Definition 6.3. In that sense, Definition 6.3 "extends" Definitions 6.2 and 6.4.*

(4) *All three definitions are equivalent when $G \neq \mathbf{0}$ and U is minimally-described (see Definition 6.7). However, this is not true when $G = \mathbf{0}$, as we will show in Section 3.2 of this chapter, causing a condition of ill-definition.*

(5) *In the remainder of this chapter, we will use the following terminologies interchangeably: "A is an EF of B"; "A is an extention of B"; "A extends B"; "B is extended by A".*

Remark 6.3. *With respect to Definition 6.2, the following alternate definition of a projection is provided by Conforti et al. (2010; 2013):*

Given a polyhedron $U = \{(x, w) \in \mathbb{R}^{p+q} : Gx + Hw \leq g\}$, its projection onto the x-space is $\varphi_x(U) = \{x \in \mathbb{R}^p : uGx \leq ug$ for all $u \in C_Q\}$, where $C_Q := \{u \in \mathbb{R}^m : uH = \mathbf{0}, u \geq \mathbf{0}\}$.

Now, assume $G = \mathbf{0}$ in this and Definition 6.2. Then, we would have:

$$\varphi_x(U) = \{x \in \mathbb{R}^p : \mathbf{0} \cdot x \leq ug \text{ for all } u \in C_Q\}$$
$$= \{x \in \mathbb{R}^p : ug \geq 0 \text{ for all } u \in C_Q\}.$$

Hence, exactly one of the following would be true:

$$\varphi_x(U) = \varnothing \text{ (if } ug < 0 \text{ for some } u \in C_Q), \text{ or}$$

$$\varphi_x(U) = \mathbb{R}^p \text{ (if } ug \geq 0 \text{ for all } u \in C_Q).$$

Hence, $\varphi_x(U)$ could not be equal to a nonempty polytope.

Definition 6.5 ("Row-redundancy"). Let $P := \{x \in \mathbb{R}^p : Ax \le a\}$, where $A \in \mathbb{R}^{m \times p}$ and $a \in \mathbb{R}^m$.

(1) We say that P has "row-redundancy" if there exists a (non-trivial) row partitioning of P with $A = \begin{bmatrix} \overline{A}_1 \\ \overline{A}_2 \end{bmatrix}$, and $a = \begin{bmatrix} \overline{a}_1 \\ \overline{a}_2 \end{bmatrix}$ (where $\overline{A}_1 \in \mathbb{R}^{n \times p}$, $\overline{A}_2 \in \mathbb{R}^{(m-n) \times p}$, $\overline{a}_1 \in \mathbb{R}^n$ and $\overline{a}_2 \in \mathbb{R}^{(m-n)}$) such that one of the following conditions is true:

(a) $(\overline{x} \in P) \Longleftrightarrow (\overline{x} \in \{x \in \mathbb{R}^p : \overline{A}_1 x \le \overline{a}_1\})$, or

(b) $(\overline{x} \in P) \Longleftrightarrow (\overline{x} \in \{x \in \mathbb{R}^p : \overline{A}_2 x \le \overline{a}_2\})$.

(2) We say that the constraints $\overline{A}_2 x \le \overline{a}_2$ are "redundant" for $\{x \in \mathbb{R}^p : \overline{A}_1 x \le \overline{a}_1\}$ if condition (1.a) is true. Similarly, we say that the constraints $\overline{A}_1 x \le \overline{a}_1$ are "redundant" for $\{x \in \mathbb{R}^p : \overline{A}_2 x \le \overline{a}_2\}$ if condition (1.b) is true.

Definition 6.6 ("Column-redundancy"). Let $P := \{x \in \mathbb{R}^p : Ax \le a\}$, where $A \in \mathbb{R}^{m \times p}$ and $a \in \mathbb{R}^m$. Let x denote the descriptive variables of P. Let $\begin{bmatrix} \overline{x}_1 \\ \overline{x}_2 \end{bmatrix}$ be a (non-trivial) partitioning of x, where $\overline{x}_1 \in \mathbb{R}^q$, and $\overline{x}_2 \in \mathbb{R}^{(p-q)}$.

(1) We say that P has "column-redundancy" if one of the following conditions is true:

(a) $\exists (B_1, b_1) \in \mathbb{R}^{n \times q} \times \mathbb{R}^n$:

$$\left(\begin{bmatrix} \overline{\overline{x}}_1 \\ \overline{\overline{x}}_2 \end{bmatrix} \in Ext(P) \Longrightarrow \overline{\overline{x}}_1 \in Ext\left(\{x \in \mathbb{R}^q : B_1 x \le b_1\}\right), \right.$$

and

$$\overline{\overline{x}}_1 \in Ext\left(\{x \in \mathbb{R}^q : B_1 x \le b_1\}\right)$$
$$\left. \Longrightarrow \exists \overline{\overline{x}}_2 \in \mathbb{R}^{(p-q)} : \begin{bmatrix} \overline{\overline{x}}_1 \\ \overline{\overline{x}}_2 \end{bmatrix} \in Ext(P) \right)$$

(where $1 \le n \le m$), or

(b) $\exists (B_2, b_2) \in \mathbb{R}^{n \times q} \times \mathbb{R}^n$:

$$\left(\begin{bmatrix} \overline{\overline{x}}_1 \\ \overline{\overline{x}}_2 \end{bmatrix} \in Ext(P) \right.$$

$$\implies \overline{\overline{x}}_2 \in Ext \left(\left\{ x \in \mathbb{R}^{(p-q)} : B_2 x \leq b_2 \right\} \right),$$

and

$$\overline{\overline{x}}_2 \in Ext \left(\left\{ x \in \mathbb{R}^{(p-q)} : B_2 x \leq b_2 \right\} \right)$$

$$\implies \exists \overline{\overline{x}}_1 \in \mathbb{R}^q : \begin{bmatrix} \overline{\overline{x}}_1 \\ \overline{\overline{x}}_2 \end{bmatrix} \in Ext(P) \right)$$

(where $1 \leq n \leq m$).

(2) We say that the variables $\overline{x}_2$ are "redundant" for $\{x \in \mathbb{R}^q : B_1 x \leq b_1\}$ when Condition (1.a) is true. Similarly, we say that variables $\overline{x}_1$ are "redundant" for $\{x \in \mathbb{R}^{(p-q)} : B_2 x \leq b_2\}$ when Condition (1.b) is true.

Definition 6.7 ("Minimally-described" polytope). We say that a polyhedron P is "minimally-described", or that (the statement of) P is "minimal", if P has no *row-redundancy* and no *column-redundancy*.

Assumption 6.1. *In the remainder of this book, with respect to Definitions 6.2–6.4, we will assume (implicitly) that U is minimally-described whenever we will be considering (or referring to) the case in which $G \neq \mathbf{0}$.*

Observation 6.2.4 and Remark 6.3 are the key points in the concept of *ill-definition* of an EF which occurs in the special case of $G = \mathbf{0}$ in Definitions 6.2–6.4. This allows for EFs *barrier* to be removed by using an alternate formulation for a given COP at hand.

3. *Ill-Definition* Condition for *EFs*

3.1. *Polytope augmentation*

Definition 6.8 ("Class of variables"). We refer to a set of variables which model a given aspect of a problem, as a "class of variables."

The *traditional* x-*variables* for example, would constitute one *class of variables* in a TSP model, as they represent (single) "travel legs" in the TSP. Similarly, the y- and z-variables used in the models of Diaby (2007) and Chapters 2–3 of this book, respectively, would constitute two distinct *classes of variables*, with the y-variables modeling doublets of "travel legs" in the TSP, and the z-variables modeling triplets of "travel legs" in the TSP.

Assumption 6.2. *In the remainder of this chapter, we will assume (without loss of generality) that a given class of variables is denoted by the same symbol in all of the models in which it is used. That is, we will assume that the same notation (whatever that may be) will be used to designate the traditional* x-*variables for example, in every model in which these variables are used.*

Definition 6.9 ("Independent spaces"). Let $x \in \mathbb{R}^p$ ($p \in \mathbb{N}_+$) and $w \in \mathbb{R}^q$ ($q \in \mathbb{N}_+$) be the vectors of descriptive variables for two polyhedra in $\mathbb{R}^p$ and $\mathbb{R}^q$, respectively.

(1) We say that x and w (or that the polyhedra) are in "independent spaces" if x and w do not have any *class of variables* in common. That is, we say that x and w are in "independent spaces" if the following conditions hold:

    (a) x cannot be partitioned as $x = \begin{pmatrix} \overline{x} \\ w \end{pmatrix}$;

    (b) w cannot be partitioned as $w = \begin{pmatrix} \overline{w} \\ x \end{pmatrix}$; and

    (c) $\forall m \in \mathbb{N}_+ : m < \min\{p, q\}$, $\nexists (\overline{x}, \overline{w}, v) \in \mathbb{R}^{(p-m)} \times \mathbb{R}^{(q-m)} \times \mathbb{R}^m : \Big(x$ and w can be respectively partitioned as $x = \begin{pmatrix} \overline{x} \\ v \end{pmatrix}$ and $w = \begin{pmatrix} \overline{w} \\ v \end{pmatrix} \Big)$, where v denotes a given *class of variables* for the problem at hand.

(2) We will say that x and w (or that the polyhedra they respectively describe) "overlap" if x and w have one or more *classes of variables* in common.

Regan and Lipton (2013) remarked that all polytopes may be viewed, in a degenerate way, as being part of one overall multi-dimensional space. The following alternate (and equivalent) definition of "independent spaces" is therefore useful in further clarifying the notion.

Definition 6.10 (Alternate definition of "Independent spaces"). Let P and Q be polytopes in $\mathbb{R}^{p+q}$, with descriptive variables $(x, y) \in \mathbb{R}^p \times \mathbb{R}^q$. We say that P and Q are in "independent spaces" iff exactly one of the following two conditions holds:

(1) $\{x \in \mathbb{R}^p : (\exists y \in \mathbb{R}^q : (x, y) \in P)\} = \mathbb{R}^p$ and $\{y \in \mathbb{R}^q : (\exists x \in \mathbb{R}^p : (x, y) \in Q)\} = \mathbb{R}^q$;
(2) $\{y \in \mathbb{R}^q : (\exists x \in \mathbb{R}^p : (x, y) \in P)\} = \mathbb{R}^q$ and $\{x \in \mathbb{R}^p : (\exists y \in \mathbb{R}^q : (x, y) \in Q)\} = \mathbb{R}^p$.

Example 6.1. *Definitions 6.9 and 6.10 can be illustrated as follows.*

- *Assume $x \in \mathbb{R}^2$ and $y \in \mathbb{R}^2$ refer to different classes of variables in a modeling context at hand.*
- *Let x and y be the descriptive variables for Polytopes P and Q respectively, with:*

$$P := \{x \in \mathbb{R}^2 : x_1 - x_2 \geq 6;\ 0 \leq x_1 \leq 6;\ 0 \leq x_2 \leq 5\};$$

$$Q := \{y \in \mathbb{R}^2 : y_1 + y_2 = 6;\ y_1 \geq 1.314;\ y_2 \geq 1.628\}.$$

- *Clearly, P and Q are in $\mathbb{R}^4$ in a degenerate sense, respectively.*
- *However:*

  - *P and Q are independent spaces according to Definition 6.9 directly;*
  - *Also, P and Q can be respectively re-written as:*
    $$\begin{cases} P' = \{(x, y) \in \mathbb{R}^2 \times \mathbb{R}^2 : \mathbf{A}x + \mathbf{0} \cdot y \leq \mathbf{a}\}, \text{ and} \\ Q' := \{(x, y) \in \mathbb{R}^2 \times \mathbb{R}^2 : \mathbf{0} \cdot x + \mathbf{B}y \leq \mathbf{b}\} \end{cases}$$

where

$$A = \begin{bmatrix} -1 & 1 \\ -1 & 0 \\ 1 & 0 \\ 0 & -1 \\ 0 & 1 \end{bmatrix}, \quad a = \begin{bmatrix} -6 \\ 0 \\ 6 \\ 0 \\ 5 \end{bmatrix},$$

$$B = \begin{bmatrix} 1 & 1 \\ -1 & -1 \\ -1 & 0 \\ 0 & -1 \end{bmatrix}, \quad b = \begin{bmatrix} 6 \\ -6 \\ -1.314 \\ -1.628 \end{bmatrix},$$

so that we have:

$$\begin{cases} \left\{ y \in \mathbb{R}^2 : \left(\exists x \in \mathbb{R}^2 : (x, y) \in P' \right) \right\} = \mathbb{R}^2, \text{ and} \\ \left\{ x \in \mathbb{R}^2 : \left(\exists y \in \mathbb{R}^2 : (x, y) \in Q' \right) \right\} = \mathbb{R}^2. \end{cases}$$

Hence, P' and Q' (and therefore, P and Q) are independent spaces according to Definition 6.10.

Definition 6.11 ("Polyhedron augmentation"). Let X be a non-empty polyhedron described in terms of variables $x \in \mathbb{R}^p$. Let $\overline{X}$ be a polyhedron the description of which consists of the constraints of X, plus additional variables and constraints that are not used in the description of X. We will say that $\overline{X}$ is an "augmentation" of X (or that $\overline{X}$ "augments" X) if the problem of optimizing any given linear function of x over X, is equivalent to the problem of optimizing that linear function over $\overline{X}$. In other words, let $x \in \mathbb{R}^p$ and $y \in \mathbb{R}^q$ be vectors of variables in *independent spaces*. Let $X := \{x \in \mathbb{R}^p : Ax \le a\} \ne \varnothing$, and $\overline{X} := \{(x, y) \in \mathbb{R}^{p+q} : Ax \le a; \ Bx + Cy \le b\} \ne \varnothing$ (where $A \in \mathbb{R}^{k \times p}$, $a \in \mathbb{R}^k$, $B \in \mathbb{R}^{m \times p}$, $C \in \mathbb{R}^{m \times q}$, and $b \in \mathbb{R}^m$). We say that $\overline{X}$ augments X if $(\forall x \in X, \exists y \in \mathbb{R}^q : (x, y) \in \overline{X})$.

Remark 6.4. *With respect to Definition 6.11:*

(1) *The additional variables and constraints of $\overline{X}$ are redundant for X (see Definitions 6.5 and 6.6);*
(2) *The optimization problem over $\overline{X}$ may not be equivalent to the optimization problem over X, if the objective function in the*

problem over $\overline{X}$ is changed from that of X to include non-zero terms of the new variables;

(3) *Every augmentation of X is an EF of X, but the converse is not true (since an EF of X need not include the constraints of X explicitly);*

(4) *The polyhedral set associated to an optimization problem is equivalent to all of its augmentations respectively, provided the expression of the objective function of the optimization problem is not changed;*

(5) *In the discussions to follow we will assume (without loss of generality) that the objective function is not changed when new variables and constraints are added to an optimization problem. Hence, in the discussions to follow, we will not distinguish between a polyhedral set and the optimization problem to which it is associated, except for where that causes confusion.*

Example 6.2. *We illustrate Definition 6.11 and Remark 6.4 as follows.*

Let:

(1) $x \in \mathbb{R}^p$ and $y \in \mathbb{R}^q$ *be variables in independent spaces;*

(2) $X := \{x \in \mathbb{R}^p : Ax \le a\}$;

(3) $L := \{(x,y) \in \mathbb{R}^{p+q} : Bx + Cy \le c\}$;

(4) $Y := \{y \in \mathbb{R}^q : Dy \le d\}$;

(5) $K_1 := \left\{ (x,y) \in \mathbb{R}^{p+q} : \begin{bmatrix} A & \mathbf{0} \\ B & C \end{bmatrix} \begin{bmatrix} x \\ y \end{bmatrix} \le \begin{bmatrix} a \\ c \end{bmatrix} \right\}$;

(6) $K_2 := \left\{ (x,y) \in \mathbb{R}^{p+q} : \begin{bmatrix} B & C \\ \mathbf{0} & D \end{bmatrix} \begin{bmatrix} x \\ y \end{bmatrix} \le \begin{bmatrix} c \\ d \end{bmatrix} \right\}$;

(7) $K_3 := \left\{ (x,y) \in \mathbb{R}^{p+q} : \begin{bmatrix} A & \mathbf{0} \\ B & C \\ \mathbf{0} & D \end{bmatrix} \begin{bmatrix} x \\ y \end{bmatrix} \le \begin{bmatrix} a \\ c \\ d \end{bmatrix} \right\}$

where: $A \in \mathbb{R}^{k \times p}$; $a \in \mathbb{R}^k$; $B \in \mathbb{R}^{m \times p}$; $C \in \mathbb{R}^{m \times q}$; $c \in \mathbb{R}^m$; $D \in \mathbb{R}^{l \times q}$, $d \in \mathbb{R}^l$.

Assume:

(8) (vii) $A \ne \mathbf{0}$, $B \ne \mathbf{0}$, $C \ne \mathbf{0}$, $D \ne \mathbf{0}$;

(9) B *cannot be partitioned as* $B = \left[\frac{A}{B}\right]$;

(10) C *cannot be partitioned as* $C = \left[\frac{\overline{C}}{D}\right]$;

(11) *The constraints of* L *are redundant for* X *and* Y.

Then:

(12) K_1 *is an augmentation of* X, *but not of* L, *nor of* Y;

(13) K_2 *is an augmentation of* Y, *but not of* L, *nor of* X;

(14) K_3 *is an augmentation of* X *and* Y, *but not of* L;

(15) K_1 *is an EF of* X, *but not of* L, *and may or may not be for* Y;

(16) K_2 *is an EF of* Y, *but not of* L, *and may or may not be for* X;

(17) K_3 *is an EF of* X *and* Y, *but not of* L;

(18) L *is not an augmentation of* X *nor of* Y;

(19) L *may or may not be an EF of* X;

(20) L *may or may not be an EF of* Y.

Remark 6.5. *In reference to the developments above:*

(1) *We will refer to the constraints of* L *as the "linking constraints" (for* X *and* Y*) in* K_3, *regardless of whether or not the constraints of* L *are redundant for* X *and* Y.

(2) *If* X *and* Y *are alternative correct abstractions of the requirements of some (same) given optimization problem, then there may or may not exist* B *and* C *such that* $((x, y) \in K_1 \implies y \in Y)$ *and* $((x, y) \in K_2 \implies x \in X)$. *This is exemplified by the Alternate TSP Polytope relative to the Standard TSP Polytope.*

(3) *If there exist* B *and* C *such that* $((x, y) \in K_1 \implies y \in Y)$ *and* $((x, y) \in K_2 \implies x \in X)$, *then it must be possible to attach meanings to* x *and* y, *so that* X *and* Y *are alternative correct abstractions of the requirements of some (same) given optimization problem. This is exemplified by the LP models of the TSP in Diaby (2007) and in Chapters 2–3 of this book, respectively, relative to the Alternate TSP Polytope, or relative to the Standard TSP Polytope.*

(4) *The main point of our developments in Section 3 below will be to show that there exists no well-defined (non-ambiguous, meaningful) EF relationship between* X *and* Y.

In particular, we will show that

$$(\exists (B,C) : (x,y) \in K_1$$

$$\implies y \in Y) \not\Rightarrow (X \text{ is a well-defined EF of } Y),$$

and that similarly,

$$(\exists (B,C) : (x,y) \in K_2$$

$$\implies x \in X) \not\Rightarrow (Y \text{ is a well-defined EF of } X).$$

For example, there exist linear transformations which allow for points of the Alternate TSP Polytope to be "retrieved" from (given) solutions of the TSP LP models in Diaby (2007) and in Chapters 2–3 of this book. Note however, that the "retrieval" of points of the Standard TSP Polytope can be accomplished only through the use of "implicit" information (about TSP node "1" specifically) that is outside the scope of the TSP LP models per se. Hence, the TSP LP models can be well-defined EFs of the Standard TSP Polytope only if they are well-defined EFs of the Alternate TSP Polytope, which would seem to suggest that the Alternate TSP Polytope must be a well-defined EF of the Standard TSP polytope. We do not believe such a suggestion is the intent of any EFs work. However, we believe the definitions of an EF must be properly interpreted in order for them not to lead to such conclusions. Specifically, using the notion of augmentation discussed in this section, we will show in the next section (Section 3) that the notion of an EF can become ill-defined (and thereby lose its meaningfulness) when the polytopes being related are expressed in coordinate systems that are independent of each other.

We will now discuss two results which will be helpful subsequently in showing the differences between the case of polytopes with *minimal descriptions* in *overlapping spaces* and the case of polytopes in *independent spaces*, as pertains to *extension* relationships.

Theorem 6.1. *Let P_1 and P_2 be non-empty, minimally-described polytopes in overlapping spaces with the set of the descriptive* *variables of P_1 included in that of P_2. An augmentation of P_2 is an EF of P_1 if and only if P_2 is an EF of P_1, according to Definitions 6.2–6.4, respectively.*

In other words, let:

$$P_1 := \{x \in \mathbb{R}^{q_1} : A_1 x \leq a_1\} \ (where \ A_1 \in \mathbb{R}^{r_1 \times q_1}; \ a_1 \in \mathbb{R}^{r_1});$$

$$P_2 := \{(x, u) \in \mathbb{R}^{q_1 + q_2} : A_2 x + Bu \leq b\}$$

(where: $A_2 \in \mathbb{R}^{r_2 \times q_1}; \ B \in \mathbb{R}^{r_2 \times q_2}; \ b \in \mathbb{R}^{r_2}$).

Assume $A_1 \neq \mathbf{0}$, $A_2 \neq \mathbf{0}$, $P_1 \neq \varnothing$, and $P_2 \neq \varnothing$. Then, an arbitrary augmentation, P_3, of P_2 is an EF of P_1 if and only if P_2 is an EF of P_1, according to Definitions 6.2–6.4, respectively.

Proof. First, note that Definitions 6.2–6.4 are equivalent to one another with respect to *extension* relations for P_1, P_2, and P_3 (see Remark 6.2.4). Hence, it is sufficient to prove the theorem for the *standard definition* (Definition 6.2).

P_3 can be written as:

$$P_3 := \{(x, u, v) \in \mathbb{R}^{q_1 + q_2 + q_3} : A_2 x + Bu \leq b; \ A_3 x + Cu + Dv \leq c\}$$

(where: $A_3 \in \mathbb{R}^{r_3 \times q_1}; \ C \in \mathbb{R}^{r_3 \times q_2}; D \in \mathbb{R}^{r_3 \times q_3}; \ c \in \mathbb{R}^{r_3}$; and $A_3 x + Cu + Dv \leq c$ are *redundant* for P_2).

$(A_3 x + Cu + Dv \leq c)$ *redundant* for $P_2 \implies$:

$$\forall (x, u) \in P_2, \quad \exists v \in \mathbb{R}^{q_3} : (x, u, v) \in P_3. \tag{6.1}$$

$(6.1) \implies$:

$$\{x \in \mathbb{R}^{q_1} : (\exists (u, v) \in \mathbb{R}^{q_1 + q_2} : (x, u, v) \in P_3)\}$$

$$= \{x \in \mathbb{R}^{q_1} : (\exists u \in \mathbb{R}^{q_2} : (x, u) \in P_2)\}. \tag{6.2}$$

$(6.2) \implies$:

$$(\varphi_x(P_3) = P_1) \iff (\varphi_x(P_2) = P_1). \qquad \square$$

We will show in the next section (Section 3.2) that Theorem 6.1 does not hold for polyhedra which are stated in *independent spaces* (such as P and Q in Example 6.1, or X and Y in Example 6.2), and that, as indicated in Remark 6.5, there cannot exist any *well-defined* (non-ambiguous, meaningful) *extension* relationship between such polytopes.

3.2. *Degeneracy condition for EFs*

Referring back to the *standard* and *alternate definitions* of an EF (i.e., Definitions 6.2–6.4, respectively), it is easy to verify (as indicated in Remark 6.2.4) that these three definitions are equivalent when $G \neq \mathbf{0}$ (with U *minimally-described*). In other words, one can easily verify that provided $G \neq \mathbf{0}$, U is an EF of X according to one of these definitions if and only if U is an EF of X according to the other definitions. However, this is not true when $G = \mathbf{0}$.

A basic intuition of Definitions 6.2–6.4 is that if the projection of U onto x-space is equal to X, then the description of X must be implicit in a constraint set of the form:

$$Gx \leq \bar{g}_w.$$

Hence, the notion of an EF can become *ill-defined* when $G = \mathbf{0}$ (i.e., when U and X are *independent spaces*). In essence, to put it roughly, there is "nothing" in the statement of U for the constraints of X to be "implicit in" (in U) when $G = \mathbf{0}$. Indeed, as we will show in the discussion to follow, when $G = \mathbf{0}$, Definition 6.3 can become contradictory of Definitions 6.2 and 6.4, resulting in an *ill-definition* (ambiguity) condition.

Theorem 6.2 shows that there exist no *extension* relations between polytopes stated in *independent spaces* according to the *standard definition* (Definition 6.2), or the *second alternate definition* (Definition 6.4) of EFs.

Theorem 6.2. *Polytopes described in independent spaces cannot be EFs of each other according to the standard definition (Definition 6.2) or the second alternate definition (Definition 6.4) of EFs.*

In other words, let $X := \{x \in \mathbb{R}^p : Ax \leq a\}$ *and* $U := \{w \in \mathbb{R}^q : Hw \leq h\}$ *be (non-empty) polytopes in independent spaces* (*where: $A \in \mathbb{R}^{m \times p}$; $a \in \mathbb{R}^m$; $H \in \mathbb{R}^{n \times q}$; $h \in \mathbb{R}^n$*). *Then:*

(i) *U cannot be an EF of X (and vice versa) according to Definition 6.2;*

(ii) *U cannot be an EF of X (and vice versa) according to Definition 6.4.*

Proof. We will show that U cannot be an EF of X according to Definitions 6.2 and 6.4, respectively. The proofs that X cannot be an EF of U according to the definitions (Definitions 6.2 and 6.4, respectively) are similar and will be therefore omitted.

Let U be re-stated in $\mathbb{R}^{p+q}$ as:

$$U' := \left\{ (x, w) \in \mathbb{R}^{p+q} : \mathbf{0} \cdot x + Hw \leq h \right\}. \tag{6.3}$$

Clearly, we have:

$$(x, w) \in U' \Longleftrightarrow w \in U. \tag{6.4}$$

Hence:

$$U \neq \varnothing \Longrightarrow U' \neq \varnothing \Longrightarrow \left(\forall x \in \mathbb{R}^p, \exists w \in \mathbb{R}^q : (x, w) \in U' \right). \tag{6.5}$$

Now consider *conditions* (i) and (ii) of the theorem. We have the following.

(i) *Condition* (1).

Using (6.4) and Definition 6.2, we have:

$$\varphi_x(U) = \varphi_x(U') = \left\{ x \in \mathbb{R}^p : \left(\exists w \in \mathbb{R}^q : (x, w) \in U' \right) \right\} = \mathbb{R}^p. \tag{6.6}$$

Because X is bounded, we must have:

$$X \subset \mathbb{R}^p. \tag{6.7}$$

Combining (6.6) and (6.7) gives:

$$\varphi_x(U) = \mathbb{R}^p \neq X. \tag{6.8}$$

(ii) *Condition* (2).

(6.5) $\Longrightarrow$:

$$\exists x \in \mathbb{R}^d \backslash X : \left(\exists w \in \mathbb{R}^q : (x, w) \in U' \right). \tag{6.9}$$

(6.4) and (6.9) $\Longrightarrow$:

$$(w \in U) \Longleftrightarrow (x, w) \in U' \not\Longleftrightarrow x \in X. \tag{6.10}$$

Hence, the *"if and only if"* condition of Definition 6.4 cannot be satisfied in general.

$\square$

Remark 6.6. *Theorem 6.2 is consistent with Remark 6.3 (p. 114).*

Corollary 6.1. *Let $X := \{x \in \mathbb{R}^p : Ax \leq a\}$ and $U := \{w \in \mathbb{R}^q : Hw \leq h\}$ be (non-empty) polytopes in independent spaces (where: $A \in \mathbb{R}^{m \times p}$; $a \in \mathbb{R}^m$; $H \in \mathbb{R}^{n \times q}$; $h \in \mathbb{R}^n$). Then, exactly one of the following is true:*

(i) *There exists no EF relationship between X and U (i.e., there exists no linear map $\pi_x : \mathbb{R}^q \longrightarrow \mathbb{R}^p$ such that $\pi_x(U) = X$, and there exists no linear map $\pi_w : \mathbb{R}^p \longrightarrow \mathbb{R}^q$ such that $\pi_w(X) = U$);*

(ii) *The EF relationship between X and U is ill-defined due to Definition 6.3 being inconsistent with Definitions 6.2 and 6.4, respectively (i.e., if there exists a linear map $\pi_x : \mathbb{R}^q \longrightarrow \mathbb{R}^p$ such that $\pi_x(U) = X$, or there exists a linear map $\pi_w : \mathbb{R}^p \longrightarrow \mathbb{R}^q$ such that $\pi_w(X) = U$, or both).*

Example 6.3. *Corollary (6.1.ii) can be illustrated using the polytopes P and Q of Example 6.1.*

We have:

(1) $\varphi_y(P) = \mathbb{R}^2 \neq Q$ *and* $\varphi_x(Q) = \mathbb{R}^2 \neq P$.

Hence, according to Definition 6.2, there exists no extension relationship between P and Q;

(2) $\exists x \notin P : (\exists y \in \mathbb{R}^2 : (x, y) \in Q)$, *which implies:* $(x \in P \not\Longleftrightarrow (\exists y \in \mathbb{R}^2 : (x, y) \in Q))$. *Similarly,* $\exists y \notin Q : (\exists x \in \mathbb{R}^2 : (x, y) \in Q)$, *which implies:* $(y \in Q \not\Longleftrightarrow (\exists x \in \mathbb{R}^2 : (x, y) \in P))$.

Hence, according to Definition 6.4, there exists no extension relationship between P and Q;

(3) $(x, y) \in (P, Q) \implies \begin{bmatrix} x_1 \\ x_2 \end{bmatrix} = \begin{bmatrix} 1 & 1 \\ 0 & 0 \end{bmatrix} \begin{bmatrix} y_1 \\ y_2 \end{bmatrix}.$

In other words, $(x, y) \in (P, Q) \implies x = Ay,$ *where* $A = \begin{bmatrix} 1 & 1 \\ 0 & 0 \end{bmatrix}$ *is the matrix for a linear transformation which maps P and Q.*

Hence, according to Definition 6.3, Q is an EF of P, which is in contradiction of (1) *and* (2) *above.*

Remark 6.7. *Part* (3) *of Example* 6.3 *shows that the polytopes of Example* 6.1 *are a counter-example to the claim whereby the existence of a linear transformation mapping the solutions of two polytopes is sufficient in order to conclude that there exists a well-defined/meaningful EF relationship (i.e., one from which valid inferences can be made) between the polytopes.*

The *ill-definition* condition stated in Corollary 6.1 will be further developed in the remainder of this section. We start with a notion which essentially generalizes the idea of the linear map (π) in Definition 6.3 with respect to the task of optimizing a linear function over a polyhedral set (since each of the linear equations which specify π must be *valid* for U and X, respectively).

Theorem 6.3. *Any two non-empty polytopes expressed in independent spaces can be respectively augmented into being EFs of each other. In other words, let* $x^1 \in \mathbb{R}^{n_1}$ ($n_1 \in \mathbb{N}_+$) *and* $x^2 \in \mathbb{R}^{n_2}$ ($n_2 \in \mathbb{N}_+$) *be vectors of variables in two independent spaces. Then, every non-empty polytope in* x^1 *can be augmented into an EF of every other non-empty polytope in* x^2, *and vice versa.*

Proof. The proof is essentially by construction.

Let P_1 and P_2 be polytopes specified as:

$$P_1 = \{x^1 \in \mathbb{R}^{n_1} : A_1 x^1 \leq a_1\} \neq \varnothing$$

$$\text{(where } A_1 \in \mathbb{R}^{p_1 \times n_1}, \text{ and } a_1 \in \mathbb{R}^{p_1});$$

$$P_2 = \{x^2 \in \mathbb{R}^{n_2} : A_2 x^2 \leq a_2\} \neq \varnothing$$

$$\text{(where } A_2 \in \mathbb{R}^{p_2 \times n_2}, \text{ and } a_2 \in \mathbb{R}^{p_2}).$$

Clearly, $\forall (x^1, x^2) \in P_1 \times P_2$, $\forall q \in \mathbb{N}_+$, $\forall B_1 \in \mathbb{R}^{q \times n_1}$, $\forall B_2 \in \mathbb{R}^{q \times n_2}$, there exists $u \in \mathbb{R}^q_{\not\prec}$ such that the constraints

$$B_1 x^1 + B_2 x^2 - u \leq 0 \qquad (6.11)$$

are *valid* for P_1 and P_2, respectively (i.e., they are *redundant* for P_1 and P_2, respectively).

Now, consider:

$$W := \left\{ (x^1, x^2, u) \in \mathbb{R}^{n_1} \times \mathbb{R}^{n_2} \times \mathbb{R}^q_{\not\prec} : \right.$$

$$C_1 A_1 x^1 \leq C_1 a_1; \qquad (6.12)$$

$$B_1 x^2 + B_2 x^1 - u \leq 0; \qquad (6.13)$$

$$\left. C_2 A_2 x^2 \leq C_2 a_2 \right\} \qquad (6.14)$$

where: $C_1 \in \mathbb{R}^{p_1 \times p_1}$ and $C_2 \in \mathbb{R}^{p_2 \times p_2}$ are diagonal matrices with non-zero diagonal entries.

Clearly, W *augments* P_1 and P_2 respectively. Hence:

$$W \text{ is equivalent to } P_1, \text{ and} \qquad (6.15)$$

$$W \text{ is equivalent to } P_2. \qquad (6.16)$$

Also clearly, we have:

$$\varphi_{x^1}(W) = P_1 \quad (\text{since } P_2 \neq \varnothing, \text{ and } ((6.13) \text{ and } (6.14)$$
$$\text{are } redundant \text{ for } P_1)), \text{ and} \qquad (6.17)$$
$$\varphi_{x^2}(W) = P_2 \quad (\text{since } P_1 \neq \varnothing, \text{ and } ((6.12) \text{ and } (6.13)$$
$$\text{are } redundant \text{ for } P_2)). \qquad (6.18)$$

It follows from the combination of (6.15) and (6.18) that P_1 is an EF of P_2.

It follows from the combination of (6.16) and (6.17) that P_2 is an EF of P_1. $\qquad\qquad\qquad\qquad\qquad\qquad\qquad\qquad \square$

Corollary 6.2. *Provided polytopes can be arbitrarily augmented for the purpose of establishing EF relationships, every two (non-empty) polytopes expressed in independent spaces are EFs of each other.*

Theorem 6.3 and Corollary 6.2 are illustrated further.

Example 6.4.

Let

$$P_1 = \{x \in \mathbb{R}^2_{\not\prec} : 2x_1 + x_2 \leq 6\};$$

$$P_2 = \{w \in \mathbb{R}^3_{\not\prec} : 18w_1 - w_2 \leq 23;\ 59w_1 + w_3 \leq 84\}.$$

For arbitrary matrices B_1, B_2, C_1, *and* C_2 *(of appropriate dimensions, respectively); say* $B_1 = \begin{bmatrix} -1 & 2 \\ 3 & -4 \end{bmatrix}$, $B_2 = \begin{bmatrix} 5 & -6 & 7 \\ -10 & 9 & -8 \end{bmatrix}$, $C_1 = [7]$, *and* $C_2 = \begin{bmatrix} 2 & 0 \\ 0 & 0.5 \end{bmatrix}$; P_1 *and* P_2 *can be augmented into EFs of each other using* $u \in \mathbb{R}^2_{\not\prec}$ *and* W:

$$W = \left\{ (x, w, u) \in \mathbb{R}^{2+3+2}_{\not\prec} : \quad [7] \begin{bmatrix} 2 & 1 \end{bmatrix} \begin{bmatrix} x_1 \\ x_2 \end{bmatrix} \leq 42; \right.$$

$$\begin{bmatrix} -1 & 2 \\ 3 & -4 \end{bmatrix} \begin{bmatrix} x_1 \\ x_2 \end{bmatrix} + \begin{bmatrix} 5 & -6 & 7 \\ -10 & 9 & -8 \end{bmatrix} \begin{bmatrix} w_1 \\ w_2 \\ w_3 \end{bmatrix} - \begin{bmatrix} u_1 \\ u_2 \end{bmatrix} \leq \begin{bmatrix} 0 \\ 0 \end{bmatrix};$$

$$\left. \begin{bmatrix} 2 & 0 \\ 0 & 0.5 \end{bmatrix} \begin{bmatrix} 18 & -1 & 0 \\ 59 & 0 & 1 \end{bmatrix} \begin{bmatrix} w_1 \\ w_2 \\ w_3 \end{bmatrix} \leq \begin{bmatrix} 46 \\ 42 \end{bmatrix} \right\}.$$

Remark 6.8.

(1) *According to Corollary 6.2, the notion of EF becomes degenerate when* $G = \mathbf{0}$ *in Definitions 6.2–6.4, respectively, and one tries to apply it by augmenting one of the polytopes at hand.*

(2) *Theorem 6.3 and Corollary 6.2 are not true for polytopes expressed in overlapping spaces, and in fact, these two results are in contradiction of Theorem 6.1. Hence, whereas one can arbitrarily augment polytopes in overlapping spaces for the purpose of establishing EFs relationships, such an approach is invalid*

(cannot produce valid results) for polytopes stated in independent spaces.

Clearly, any notion of "extension" which allows for an object to be extensions of its own extensions cannot be a well-defined one (i.e., must be an ill-defined one), unless the objects involved are indistinguishable from their respective "extensions". For example, clearly, one cannot reasonably argue that P_1 and P_2 in Example 6.4 above are EFs of each other in a meaningful sense.

4. Redundancy Matters for Polytopes Stated in *Independent Spaces*

The notion of *independent spaces* we have introduced in this chapter is important because, as we have shown, it refines the notion of EFs by separating the case in which that notion is degenerate (with every polytope potentially being an EF of every other polytope) from the case in which the notion of EF is *well-defined*/meaningful. It separates the case in which the addition of *redundant* constraints and variables (for the purpose of establishing EFs relations) matters (i.e., makes a difference to the outcome of analysis) from the case in which the addition of *redundant* constraints and variables does not matter.

Two key results of Section 3 of this chapter are that:

(1) If U in $\mathbb{R}^p$ and V in $\mathbb{R}^q$ are in *overlapping spaces*, then an *augmentation* of V is an EF of U if and only if V is an EF of U. (This is the case in which the addition of *redundant* constraints does not matter, and is stated in Theorem 6.1.)

(2) But (1) is not true if U and V are in *independent spaces*. As we have shown in Section 3.2, if polytopes can always be arbitrarily augmented for the purpose of establishing EFs relations, then any two polytopes that are in *independent spaces* are EFs of each other. (This is the case in which adding *redundant* constraints and variables does matter, and is stated in Theorem 6.3).

Because of (2), the addition of *redundant* constraints and variables for the purpose of establishing EFs relations can lead to ambiguities when applied to polytopes in *independent spaces*. This ambiguity would stem from the fact that one would reach contradicting conclusions depending on what we do with the *redundant* constraints and variables which are introduced in order to make the models *overlap*. To clarify this: Assume V is augmented with the variables of U plus the constraints for the linear transformation that establishes the one-to-one correspondence between U and V; call this augmented V, V'. In EFs work, it is commonly assumed/suggested that *redundant* constraints and variables of a model can be removed from it without any loss of generality. In the case of V', this would lead to contradictory conclusions with respect to the question of whether or not V' is an EF of U. If the added *redundant* constraints and variables are kept, the answer would be "yes"; if they are removed, V' would "revert" back to V, so that the answer would be "no". This is the inconsistency which is pointed out in Theorem 6.2 and Corollary 6.1, and illustrated in Example 6.4.

A specific implication of (2) is that, provided polytopes can be arbitrarily augmented for the purpose of EFs, every conceivable polytope that is non-empty and does not require the *traditional x-variables* (i.e., the city-to-city $x_{i,j}$ variables) of the *Standard TSP Polytope*; see Definitions 1.2 and 6.1 in its description is an EF of the *Standard TSP Polytope*, and vice versa. Clearly, this can only be in a degenerate/non-meaningful sense from which no valid inferences can be made.

4.1. *Alternate/auxiliary models*

We will now provide some insights into the meaning/consequence of the existence of an affine map establishing a one-to-one correspondence between polytopes that are stated in *independent spaces*, as brought to our attention in private discussions by Yannakakis (2013). The linear map stipulated in Definition 6.3 is a special case of the affine map. Referring back to Definitions 6.2–6.4, assume $G = \mathbf{0}$ in the expression of U. We will show that in that case, provided the constraints expressing the affine transformation are *redundant* for U, then U is simply an alternate model (a "reformulation") of the underlying problem at hand, and can be used in an "auxiliary" way, in order to solve the optimization problem over P without any reference to/knowledge of the $\mathcal{H}$-description of P (see Remark 2.1). In other words, if $G = \mathbf{0}$ in Definitions 6.2–6.4 and there exists an affine transformation which performs a one-to-one map of U onto P and is *redundant* for P and U respectively, then the optimization of any linear function over P can be done without any regard to the $\mathcal{H}$-description of P, by resorting to Q. This is more detailed in the following discussion.

Remark 6.9.

- *Referring back to Example 6.2 (p. 120), assume that the non-negativity requirements for x and y are included in the constraints of X and Y, respectively, and that L has the form:*

$$L = \{(x, y) \in \mathbb{R}^{p+q}_{\not{\star}} : x - Cy = b\} \quad (\text{where } C \in \mathbb{R}^{p \times q}, \text{ and } b \in \mathbb{R}^{p}).$$

 (Recall that the constraints of L are assumed to be redundant for X and Y, respectively.)
- *Consider the optimization problem:*

  Problem LP_1:

Minimize: $\alpha^T x,$
Subject to: $(x, y) \in L; \ \ y \in Y$
(where $\alpha \in \mathbb{R}^p$).

- *Problem LP_1 is equivalent to the smaller linear program:*

  Problem LP_2:

Minimize: $\left(\alpha^T C\right) y + \alpha^T b,$
Subject to: $y \in Y,$
(where $\alpha \in \mathbb{R}^p$).

 Hence, if L is the graph of a one-to-one correspondence between the points of X and the points of Y (see Beachy and Blair (2006, *pp. 47–59)), then, the optimization of any linear function of x over X can be done by first using Problem LP_2 in order to get an optimal y, and then using Graph L to "retrieve" the corresponding x. Note that the second term of the objective function of Problem LP_2 can be ignored in the optimization process of Problem LP_2, since that term is a constant.*

Hence, if L is derived from knowledge of the $\mathcal{V}$-representation of X only (as is the case for the TSP LP models of Diaby (2007) and Chapters 2–3 of this book relative to the Standard TSP Polytope, for example), then this would mean that the $\mathcal{H}$-representation of X is not involved in the "two-step" solution process (of using Problem LP_2 and then Graph L), but rather, that only the $\mathcal{V}$-representation of X is involved (see Remark 2.1).

Hence, in general, when $G = 0$ in Definition 6.3, the condition stipulated in that definition cannot imply (or lead to) EFs relations which are meaningful in relating the minimal $\mathcal{H}$-representations of the polytopes involved (see Examples 6.1 and 6.3 also).

Direct corollaries of the developments above in this section and in Section 3 are the following.

Corollary 6.3. *Let P and Q be (non-empty) polytopes stated in overlapping spaces. Assume (without loss of generality) that the set of the descriptive variables of P is embedded in the set of the descriptive variables of Q. If the descriptive variables of P are redundant in Q (i.e., if all of the constraints involving the variables of P can be dropped from Q after the variables of P are substituted out of the objective function of the optimization problem over Q; see Definition 6.6 (p. 115)), then Q **can be** expressed in independent space relative to P. Hence, Q would not be (and could not be augmented into) a well-defined, non-degenerate, non-ambiguous, meaningful EF of P.*

Corollary 6.4. *EFs developments relating problem sizes (such as Yannakakis (1991), and Fiorini et al. (2011; 2012), in particular) are valid/applicable only when the projections involved are irredundant-component projections (i.e., when the variables being projected to are not redundant (see Definition 6.6) in the model at hand).*

4.2. *The case of the minimum spanning tree problem*

A case in point for the discussions in Section 4.1 is that of MSTP. We believe the refinement brought to the notion of EFs by our notion of *independent spaces* provides some understanding as to why EFs "barriers" approaches do not apply to Edmond's formulation of the MSTP (beyond the awareness of the fact that polynomial models and procedures have been discovered for the MSTP). This is developed in the following discussion.

Example 6.5. *We show that Martin's polynomial-sized LP model of the MSTP is not an EF (in a non-degenerate, meaningful sense) of Edmonds's exponential LP model of the MSTP, by showing that Martin's model can be stated in independent space relative to Edmonds' model.*

- **Using the notation in Martin (1991), i.e.,**

  — $N := \{1, \ldots, n\}$ (Set of vertices);
  — E : Set of edges;
  — $\forall S \subseteq N, \gamma(S)$: Set of edges with both ends in S.

- **Exponential-sized/"sub-tour elimination" LP formulation (Edmonds (1970))**

  (P):

  Minimize: $\sum_{e \in E} c_e x_e,$

  Subject to: $\sum_{e \in E} x_e = n - 1;$

  $\sum_{e \in \gamma(S)} x_e \leq |S| - 1; \quad S \subset E;$

  $x_e \geq 0 \ \text{ for all } e \in E.$

- **Polynomial-sized LP reformulation (Martin (1991))**

  (Q):

  Minimize: $\sum_{e \in E} c_e x_e,$

  Subject to: $\sum_{e \in E} x_e = n - 1;$

  $$z_{k,i,j} + z_{k,j,i} = x_e; \quad k = 1, \ldots, n; \quad e \in \gamma(\{i, j\});$$

  $$\sum_{s > i} z_{k,i,s} + \sum_{h < i} z_{k,i,h} \leq 1; \quad k = 1, \ldots, n; \quad i \neq k;$$

  $$\sum_{s > k} z_{k,k,s} + \sum_{h < k} z_{k,k,h} \leq 0; \quad k = 1, \ldots, n;$$

  $$x_e \geq 0 \ \text{ for all } e \in E; \quad z_{k,i,j} \geq 0 \ \text{ for all } k, i, j.$$

- **Restatement of Martin's LP model (Diaby and Karwan (2015); Regan (2015))**

  For each $e \in E$:

  — Denote the ends of e as i_e and j_e, respectively;

  — Fix an arbitrary node, r_e, which is not incident on e (i.e., r_e is such that it is not an end of e).

  Then, one can verify that Q is equivalent to:

  $(Q\prime)$:

  Minimize: $\sum_{e \in E} c_e z_{r_e, i_e, j_e} + \sum_{e \in E} c_e z_{r_e, j_e, i_e},$

  Subject to: $\sum_{e \in E} z_{r_e, i_e, j_e} + \sum_{e \in E} z_{r_e, j_e, i_e} = n - 1;$

  $$z_{k, i_e, j_e} + z_{k, j_e, i_e} = z_{r_e, i_e, j_e} + z_{r_e, j_e, i_e}; \quad k = 1, \ldots, n;$$
  $$e \in E;$$

$$\sum_{s>i} z_{k,i,s} + \sum_{h<i} z_{k,i,h} \leq 1; \quad i, k = 1, \ldots, n : i \neq k;$$

$$\sum_{s>k} z_{k,k,s} + \sum_{h<k} z_{k,k,h} \leq 0; \quad k = 1, \ldots, n;$$

$$z_{k,i,j} \geq 0 \quad \text{for all } k, i, j.$$

Remark 6.10.

(1) *Although solutions of P can be "retrieved" from those of Q', that "retrieval" does not require knowledge of the $\mathcal{H}$-description of P, and does not, therefore, imply any meaningful extension relationships between the $\mathcal{H}$-descriptions of P and Q' (or equivalently, between the $\mathcal{H}$-descriptions of P and Q).*

(2) *We would argue that the reason that EFs "barriers" do not apply to the MSTP (beyond the awareness that polynomial procedures have been discovered for it) is the fact that Martin's formulation of the MSTP (Q in Example 6.5) is not an EF of Edmonds' model (P in Example 6.5) in a well-defined, non-degenerate, meaningful sense. In fact, the projection of Q' onto the space of the variables of P is $\varphi_x(Q') = \mathbb{R}^{|E|} \neq P$, whereas Q' and Q are equivalent.*

4.3. *Application to the Fiorini et al. (2011; 2012) "barriers"*

We will now return our focus to the TSP in this section. We will illustrate Corollaries 6.3 and 6.4 using the developments in Fiorini *et al.* (2011; 2012), by showing that the mathematics in those papers actually "breaks down" as one tries to apply their developments when the polytopes involved are stated in *independent spaces* (i.e., when $G = \mathbf{0}$ in Definitions 6.2–6.4, respectively). As we indicated in the introduction section to this chapter, we are not concerned in this chapter with the issue of correctness/incorrectness of any particular LP model that may have been proposed for NP-Complete problems. Rather, our aim is to show that the resolution of that issue (of correctness/incorrectness) can be beyond the scope of EFs work

under some conditions, such as is the case for the LP models of
Diaby (2007), and Chapters 2–3 of this book, for example. In order
to simplify the discussion, we will focus on the *Standard TSP Poly-
tope*, and use the *Alternate TSP Polytope* discussed in Chapter 2 (see
Definition 2.1), as well as the TSP LP models of Diaby (2007) and
Chapters 2–3 of this book, respectively, as illustrations.

Fiorini *et al.* (2012) is a reorganized and extended version of
Fiorini *et al.* (2011). The key extension is the addition of another
alternate definition of an EF (Fiorini *et al.* (2012), p. 96) which we
recall in this chapter as Definition 6.4. This new alternate definition is then used to re-arrange "Section 5" of Fiorini *et al.* (2011)
into "Section 2" and "Section 3" of Fiorini *et al.* (2012). Hence, the
developments in "Section 5" of Fiorini *et al.* (2011) which depended
on "Theorem 4" of that paper, are "stand-alones" (as "Section 3")
in Fiorini *et al.* (2012), and "Theorem 4" in Fiorini *et al.* (2011) is
relabeled as "Theorem 13" in Fiorini *et al.* (2012).

Our discussion of specifics why neither of the two papers are
applicable when relating polytopes in *independent spaces* will be
based on the non-validity of the proofs of "Theorem 4" of Fiorini
et al. (2011) (which is "Theorem 13" of Fiorini *et al.* (2012), as indicated in this section), and of "Theorem 3" of Fiorini *et al.* (2012)
(which is in "Section 3" of that paper) when $G = 0$ in Definitions
6.2, 6.3, and 6.4, respectively.

Theorem 6.4. *Let $W \subset \mathbb{R}^\xi$ be the polytope involved in an arbitrary
abstraction of TSP tours. Assume W and the Standard TSP Polytope
are expressed in independent spaces (such as is the case for the Alternate TSP Polytope, or the polytopes associated with the LP models of
the TSP proposed in Diaby (2007) and in Chapters 2–3 of this book,
respectively). Then, the developments in Fiorini et al. (2011) are not
valid (and therefore, not applicable) for relating the size of W to the
size of the Standard TSP Polytope.*

Proof. Using the terminology and notation of Fiorini *et al.* (2011),
the main results of Section 2 of Fiorini *et al.* (2011) are developed in
terms of $Q := \{(x,y) \in \mathbb{R}^{d+k} \mid Ex + Fy = g, \ y \in C\}$ and $P := \{x \in \mathbb{R}^d \mid Ax \le b\}$.

Note that letting Q (in Fiorini *et al.* (2011)) stand for W, and P (in Fiorini *et al.* (2011)) stand for the *Standard TSP Polytope* respectively, E would be equal to $\mathbf{0}$ in the expression of Q. Hence, firstly, assume $E = \mathbf{0}$ in the expression of Q (i.e., $Q := \{(x, y) \in \mathbb{R}^{d+k} \mid \mathbf{0}x + Fy = g,\ y \in C\}$). Then, secondly, consider Theorem 4 of Fiorini *et al.* (2011) (which is pivotal in that work). We have the following:

(1) (i) If $A \neq \mathbf{0}$ in the expression of P, then the proof of the theorem is invalid since that proof requires setting "$E := A$" (see Fiorini *et al.* (2011, p. 7));

(2) (ii) If $A = \mathbf{0}$, then $P := \{x \in \mathbb{R}^d \mid \mathbf{0}x \leq b\}$. This implies that either $P = \mathbb{R}^d$ (if $b \geq \mathbf{0}$) or $P = \varnothing$ (if $b \not\geq \mathbf{0}$). Hence, P would be either unbounded or empty. Hence, there could not exist a polytope, $Conv(V)$, such that $P = Conv(V)$ (see Bazaraa *et al.* (2006, pp. 39–49), or Fiorini *et al.* (2011, 16–17), among others). Hence, the conditions in the statement of Theorem 4 of Fiorini *et al.* (2011) would be *ill-defined*/impossible.

Hence, the developments in Fiorini *et al.* (2011) are not applicable for W. $\qquad\square$

Theorem 6.5. *Let $W \subset \mathbb{R}^\xi$ be the polytope involved in an arbitrary abstraction of TSP tours. Assume W and the Standard TSP Polytope are expressed in independent spaces (such as is the case for the Alternate TSP Polytope, or the polytopes associated with the LP models of the TSP proposed in Diaby (2007) and Chapters 2–3 of this book, respectively). Then, the developments in Fiorini et al. (2012) are not valid (and therefore, not applicable) for relating the size of W to the size of the Standard TSP Polytope.*

Proof. First, note that "Theorem 13" of Fiorini *et al.* (2012, p. 101) is the same as "Theorem 4" of Fiorini *et al.* (2011). Hence, the proof of Theorem 6.4 is applicable to "Theorem 13" of Fiorini *et al.* (2012). Hence, the developments in Fiorini *et al.* (2012) that hinge on this result (namely, from "Section 4" of the paper, onward) are not applicable to W.

Now consider "Theorem 3" of Fiorini *et al.* (2012, Section 3, p. 99). The proof of this theorem hinges on the statement that (using the terminology and notation of Fiorini *et al.* (2012)):

$$Ax \leq b \iff \exists y : E^{\leq}x + F^{\leq}y \leq g^{\leq}, \ E^{=}x + F^{=}y \leq g^{=}. \quad (6.19)$$

Now, observe that if x and y do not *overlap*,

$$(\exists y : \mathbf{0} \cdot x + F^{\leq}y \leq g^{\leq}, \ \mathbf{0} \cdot x + F^{=}y \leq g^{=})$$

cannot imply $(Ax \leq b)$ in general.

Hence, provided x and y do not *overlap* (i.e., provided x and y are in *independent spaces*), the "if and only if" stipulation of (6.19) cannot be satisfied in general. Hence, Theorem 3 of Fiorini *et al.* (2012) is not applicable for W. □

Chapter 7

Illustrations for Other NP-Complete COPs

1. Introduction

In this chapter, we will elaborate on the COPs other than the TSP discussed in Chapter 4, namely, the Set Partitioning Problem, the Vertex Coloring Problem, and the Multiple Traveling Salesman Problem. Using the modeling principle developed in Chapters 2–4, the LP models for each of these COPs (SPP, VCP, mTSP) will be explicitly stated. Proofs of stated results will be omitted, as these can be found in Chapters 2–3 and in Diaby (2010a; 2010b; 2010c).

2. The Set Partitioning Problem (SPP)

2.1. *Introduction*

Along with the Traveling Salesmen Problem (see Chapter 1 of this book) and the Set Covering Problem (see Kinney *et al.* (2007)), the SPP (Balas and Padberg (1976)) is one of the three most applied combinatorial optimization problems. The most abundant area of application of the problem has been the transportation industry, where it has been used (in the recent literature) in contexts of dispatching vehicles (Desaulniers *et al.* (2003); Westphal and Krumke (2008)), determining vehicle fleet mixes (Lee *et al.* (2008)), routing and scheduling of vehicles (Alvarenga *et al.* (2007); Baldacci *et al.* (2008); Freling *et al.* (2003); Hong *et al.* (2009); Ileri *et al.* (2006); Jepsen *et al.* (2008); Kliewer *et al.* (2006)), scheduling of transportation system crews (Medard and Sawhney (2007); Mesquita and Paias (2008)), among many others of the recent developments. In the recent literature, outside of the transportation industry, the SPP has been used in contexts of cellular manufacturing system design (Mahdavi *et al.* (2006)), communication system network design (Oliveira *et al.* (2005); Tombus and Bilgic (2004)), computer hardware and software designs (Osei-Bryson and Joseph (2006); Thomadsen and Larsen (2007)), design of balanced student teams (Desrosiers *et al.* (2005)), design of maintenance schedules for production machines (Grigoriev *et al.* (2006)), facility location (Berger *et al.* (2007)), image compression (Jyotheswar and Mahapatra (2007)), molten iron allocation in the steel industry (Tang *et al.* (2007)), supply chain design (Chiang and Russell (2004); Sadler and Gervet (2008); Sindhuchao *et al.* (2005); Teo and Shu (2004)), "surgical theater" planning involving room and surgical team assignments in hospitals (Fei *et al.* (2008)), and workforce scheduling (Beliën and Demeulemeester (2007); Eveborn *et al.* (2006); Everborn and Ronnqvist (2004)). In this section, we will present a linear programming (LP) model of SPP using our earlier results, and illustrate it with a numerical example.

Because of the extremely broad range of applicability (as highlighted above), the SPP can be interpreted from a wide variety of perspectives. We will adopt a manufacturing/operations perspective which can be described as follows. There is a set $\Gamma = \{1, \ldots, \gamma\}$ of processors, and a set $\Theta = \{1, \ldots, \theta\}$ of tasks which must be performed using those processors. A fixed cost, c_p ($p \in \Gamma$), is incurred if processor p is used. The problem is to select a subset of the processors to be used so that each task can be performed on exactly one of the chosen processors, and the total cost of the chosen processors is minimized.

Let o denote the $\gamma \times \theta$ input matrix for the SPP, with the (p, t)th entry (denoted o_{pt}) being a 0/1 binary indicator that is equal to "1" iff task $t \in \Theta$ can be performed on processor $p \in \Gamma$. Let u_p be a 0/1 binary variable that is equal to "1" iff $p \in \Gamma$ is used. Then, the classical Integer Programming (IP) formulation of the SPP is as follows:

Problem 7.1 (Problem SPP).

   minimize:

$$\zeta(u) := \sum_{p \in \Gamma} c_p u_p, \qquad (7.1)$$

subject to:

$$\sum_{p \in \Gamma} o_{pt} u_p = 1; \quad t \in \Theta, \tag{7.2}$$

$$u_p \in \{0, 1\}; \quad p \in \Gamma, \tag{7.3}$$

where:

$$o_{pt} = \begin{cases} 1 & \text{if task } t \text{ can be performed on processor } p; \\ 0 & \text{otherwise,} \end{cases}$$

$$u_p = \begin{cases} 1 & \text{if processor } p \text{ is used;} \\ 0 & \text{otherwise,} \end{cases}$$

$c_p = \text{cost of using processor } p.$

Problem SPP was one of the first problems to be classified as *NP-Complete* (Karp, 1972). Hence, research directed at developing solution methods has focused on heuristics and efficient enumerative search procedures. Reviews can be found in Balas and Padberg (1976), Boschetti *et al.* (2008), and Joseph (2002). The LP relaxation tends to yield good ("tight") lower bounds. However, solving it can be challenging because of its high degree of degeneracy (Barahona and Anbil (2002)). Hence, heuristic precedures aimed at solving the LP relaxation have been developed (Barahona and Anbil (2002); Boschetti *et al.* (2008); Cavalcante *et al.* (2008); Chan and Yano (1992); Conforti *et al.* (2007); Fisher and Kedia (1990); Klabjan (2004); Lucena (2005)). Meta-heuristic approaches (Alvarenga *et al.* (2007); Lee *et al.* (2008)), and heuristics based on reformulations (Ali and Han (1998); Ali and Thiagarajan (1989); El-Darzi and Mitra (1992; 1995); Lewis *et al.* (2008); Sherali and Lee (1996)) have also been developed for the overall problem. The exact procedures that have been proposed have been, for the most part, enumerative search procedures (Balas and Padberg (1976); Boschetti *et al.* (2008); Chan and Yano (1992); Fisher and Kedia (1990); Harche and Thompson (1994); Hoffman and Padberg (1993); Joseph (2002); Linderoth *et al.* (2001); Marsten (1974); Marsten and Shepardson (1981)). Cutting planes methods have been developed also. These are reviewed in Balas and Padberg (1976).

Using our path-based modeling approach described in the previous chapters, we will now develop a LP formulation of the SPP. First, we will discuss a Transportation Problem (TP)-based reformulation in Section 2.2. Then, we discuss a path-based reformulation of this TP-based model in Section 2.3. The overall LP model is discussed in Section 2.4. Conclusions are discussed in Section 2.5.

2.2. *TP-based reformulation*

Unlike previous network flow-based models which have been proposed for the SPP (see Ali and Han, 1998; Ali and Thiagarajan, 1989; El-Darzi and Mitra, 1992; 1995, for example), our TP-based reformulation of *Problem SPP* does not depend on the presence of any special structures in the SPP input matrix. Hence, our TP-based reformulation is applicable to any instance of the SPP, and does not require any searches of the SPP input matrix for structures. In this section, we present the TP-based model and illustrate it with a numerical example.

Notation 7.1 (TP form notation).

(1) γ: Number of processors;
(2) θ: Number of tasks;
(3) $\Gamma := \{1, \ldots, \gamma\}$ (Set of processors);
(4) $\Theta := \{1, \ldots, \theta\}$ (Set of tasks);
(5) $T_p := \{t \in \Theta : o_{pt} = 1\}$ $\forall p \in \Gamma$ (Set of tasks that can be performed using processor p);
(6) $\tau_p := |T_p|$ $\forall p \in \Gamma$;
(7) $K_p := \{1, \ldots, \tau_p\}$ $\forall p \in \Gamma$;
(8) $n := \sum_{p \in \Gamma} \tau_p$ (Number of non-zero entries of the SPP input matrix);
(9) $P_t := \{p \in \Gamma : o_{pt} = 1\}$ $\forall t \in \Theta$ (Set of processors that can perform task t);
(10) $\pi_t := |P_t|$ $\forall t \in \Theta$;
(11) v_{pt}: 0/1 variable equal to "1" iff task t is performed using processor p.

Assumption 7.1. *We assume without loss of generality that:*

(1) The members of T_p ($p \in \Gamma$) have been arranged in increasing order of task indices, with α_{pk} ($k \in K_p$) as the index of the kth member; that is, the ordering of T_p is such that $\alpha_{pk} < \alpha_{p\ell}$ $\forall (k, \ell) \in K_p^2 : k < \ell$.

(2) A "dummy" task, $\theta + 1$, has been added to the set of tasks, with $o_{p,\theta+1} = 1$ $\forall p \in \Gamma$, and no restriction on the number of times $\theta + 1$ can be included in any SPP solution.

(3) The feasible set of *Problem SPP* is non-empty.

Definition 7.1. For $p \in \Gamma$, we refer to the members of K_p, as the "(processing) slots" on p.

Our proposed TP-based reformulation is as follows:

Problem 7.2 (*Problem SPPTP*).

minimize:

$$\zeta(v) := \sum_{p \in \Gamma} c_p v_{p, \alpha_{p,1}}, \tag{7.4}$$

subject to:

$$\sum_{p \in P_t} v_{pt} = 1; \quad t \in \Theta, \tag{7.5}$$

$$\sum_{p \in \Gamma} v_{p, \theta+1} = n - \theta; \tag{7.6}$$

$$\sum_{t \in T_p \cup \{\theta+1\}} v_{pt} = \tau_p; \quad p \in \Gamma, \tag{7.7}$$

$$v_{p, \alpha_{pk}} - v_{p, \alpha_{p,1}} = 0 \quad p \in \Gamma; \ k \in K_p \backslash \{1\}, \tag{7.8}$$

$$v_{pt} \in \{0, 1\}, \ p \in \Gamma, \ t \in T_p; \ v_{p, \theta+1} \geq 0, \ p \in \Gamma. \tag{7.9}$$

Constraints (7.5) ensure (in light of constraints (7.9)) that each task can be performed on exactly one of the chosen processors. Constraints (7.7) stipulate that every *slot* on a given processor must be assigned. Constraints (7.8) ensure that the *slots* are all assigned either to actual tasks (i.e., the processor is used), or to the *dummy task* (i.e., the processor is not used). Hence, the objective function (7.4) correctly accounts for the total cost of the processors used.

Theorem 7.1. *The following are true*:

(i) There is a one-to-one correspondence between feasible solutions of *Problem SPPTP* and SPP solutions.

(ii) There is a one-to-one correspondence between feasible solutions of *Problem SPPTP* and feasible solutions of *Problem SPP*.

(iii) *Problem SPPTP* and *Problem SPP* are equivalent optimization problems.

A numerical illustration of the *TP-based reformulation* is shown on Figure 7.1.

2.3. Path-based reformulation of the TP-based model constraints

2.3.1. *Multipartite graph representation*

We reformulate constraints (7.5)–(7.9) of *Problem TP* in terms of flows over the multipartite digraph illustrated in Figure 7.2. We refer to this graph as the "Set Partitioning Problem Flow Graph (SPPFG)". In the SPPFG, each node corresponds to a feasible (task, processor, *slot*) triplet. That is, a triplet $(t, p, k) \in (\Theta, \Gamma, K_p)$ has a corresponding node in the graph iff $t = \alpha_{pk}$. On the other hand, there is a node in the graph corresponding to each triplet $(\theta + 1, p, k)$ $(p \in \Gamma, k \in K_p)$. The arcs of the graph are specified through the explicit statements of the forward and backward stars of each of the nodes of the graph.

Definition 7.2.

(1) We refer to the set of nodes of the *SPPFG* that correspond to a given (processor, *slot*) pair as a *stage* of the graph.

(2) We refer to the set of nodes of the *SPPFG* that correspond to a given task as a *level* of the graph.

In order to simplify the exposition and to conform to the formalisms of the previous chapters, we perform a sequential indexing of the stages, as described below.

Number of processors, $\gamma = 8$;
Set of processors, $\Gamma = \{1,2,3,4,5,6,7,8\}$;
Number of tasks, $\theta = 6$;
Set of tasks, $\Theta = \{1,2,3,4,5,6\}$;

Processors

p	T_p	τ_p	c_p
1	$\{1,3,5\}$	3	4
2	$\{2,6\}$	2	3
3	$\{2\}$	1	3
4	$\{4\}$	1	2
5	$\{1,2,6\}$	3	3
6	$\{1,3\}$	2	2
7	$\{1,4,5\}$	3	3
8	$\{2,5\}$	2	4

Tasks

t	P_t	π_t
1	$\{1,5,6,7\}$	4
2	$\{2,3,5,8\}$	4
3	$\{1,6\}$	2
4	$\{4,7\}$	2
5	$\{1,7,8\}$	3
6	$\{2,5\}$	2

Ordering

p	k	α_{pk}
1	1	1
	2	3
	3	5
2	1	2
	2	6
3	1	2
4	1	4
5	1	1
	2	2
	3	6
6	1	1
	2	3
7	1	1
	2	4
	3	5
8	1	2
	2	5

(a) Problem Input Data and Notation

Tasks		Processors (Γ)								"Demand"
		p=1	p=2	p=3	p=4	p=5	p=6	p=7	p=8	
Θ	t=1	1				1	1	1		1
	t=2		1	1		1			1	1
	t=3	1					1			1
	t=4				1			1		1
	t=5	1						1	1	1
	t=6		1			1				1
$\{\theta+1\}$	t=7	1	1	1	1	1	1	1	1	11
"Supply"		3	2	1	1	3	2	3	2	17

(b) Illustration of the Transportation Tableau Form

Figure 7.1. Numerical illustration of the transportation problem-based reformulation of the SPP.

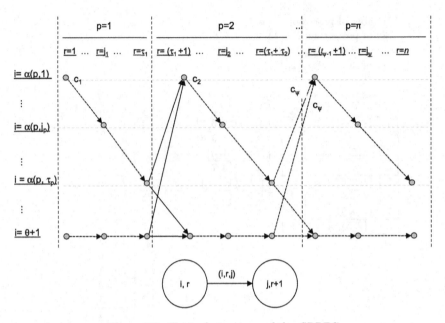

Figure 7.2. General structure of the *SPPFG*.

Notation 7.2 (SPPFG notation).

(1) $M := \Theta \cup \{\theta + 1\}$;

(2) $S := \{1, \ldots, n\}$ (Set of *stages* of the *SPPFG*);

(3) $R := S \backslash \{n\}$;

(4) $f_p := \begin{cases} 1; & \text{for } p = 1 \\ \displaystyle\sum_{j=1}^{p-1} \tau_j + 1; & \text{for } p > 1 \end{cases}$

$\forall p \in \Gamma$ (Index of the *stage* corresponding to the (processor, slot) pair, $(p, 1)$);

(5) $\ell_p := \sum_{j=1}^{p} \tau_j \ \forall p \in \Gamma$ (Index of the *stage* corresponding to the (processor, slot) pair, (p, τ_p));

(6) $\xi_r := \arg\max\{f_p : f_p \leq r\} \ \forall r \in S$ (Processor index for *stage* r);

(7) $M_r := \{\alpha_{\xi_r, r - f_{\xi_r} + 1}\} \ \forall r \in S$ (Set of tasks in Θ which define nodes at *stage* r);

(8) $N_r := M_r \cup \{\theta + 1\} = \{\alpha_{\xi_r, r - f_{\xi_r} + 1}, \theta + 1\} \ \forall r \in S$;

(9) $\overline{N} := \{[i, r], r \in S, i \in N_r\}$ (Set of nodes of the *SPPFG*);

(10)

$$
F_r(t) := \begin{cases}
N_{r+1}\backslash\{t\} & \text{for } r < n;\ r = \ell_{\xi_r};\ t \neq \theta + 1 \\[4pt]
N_{r+1} & \text{for } r < n;\ r = \ell_{\xi_r};\ t = \theta + 1 \\[4pt]
\begin{aligned}M_{r+1} \\ = \{\alpha_{\xi_r,\ r-f_{\xi_r}+2}\}\end{aligned} & \text{for } r < n;\ f_{\xi_r} \leq r < \ell_{\xi_r}; \\[2pt]
& \qquad t \neq \theta + 1 \\[4pt]
\{\theta + 1\} & \text{for } r < n;\ f_{\xi_r} \leq r < \ell_{\xi_r}; \\[2pt]
& \qquad t = \theta + 1 \\[4pt]
\varnothing & \text{for } r = n
\end{cases}
$$

$\forall r \in S,\ \forall t \in (T_{\xi_r} \cup \{\theta + 1\})$ (Forward star of node (t, r) of SPPFG);

(11) $B_r(t) := \begin{cases} \{j \in N_{r-1} : t \in F_{r-1}(j)\} & \text{for } r > 1 \\ \varnothing & \text{for } r = 1 \end{cases}$

$\forall r \in S,\ \forall t \in N_r$ (Backward star of node (t, r) of the *SPPFG*);

(12) $A := \{[i, r, j] \in (M, R, M) : i \in N_r;\ j \in F_r(i)\}$ (Set of arcs of the *SPPFG*).

The notation and structure of the *SPPFG* are illustrated in Figure 7.3 for the numerical example shown in Figure 7.1.

Remark 7.1.

(1) Each stage of the *SPPFG* comprises exactly two nodes of the graph.
(2) The maximum number of arcs originating from any *stage* of the *SPPFG* is four.

Definition 7.3. We refer to a path of the *SPPFG* which spans the set of *stages* of the *SPPFG* and which includes each task in Θ exactly once as a *SPP path* of the graph; that is, a set of arcs,

$$([i_1, 1, i_2], [i_2, 2, i_3], \ldots, [i_{n-1}, n-1, i_n]) \in A^{n-1},$$

is a *SPP path* iff $(\forall t \in \Theta,\ \exists p \in S : i_p = t)$, and $\big(\forall (i_p, i_q) \in \Theta^2,\ \forall (p, q) \in (S, S\backslash\{p\}),\ i_p \neq i_q\big)$.

$n = \sum_{p\in\Pi} \tau_p = 17;$

$S = \{1,\ldots,17\};$

$R = \{1,\ldots,16\}$

p	f_p	ℓ_p
1	1	3
2	4	5
3	6	6
4	7	7
5	8	10
6	11	12
7	13	15
8	16	17

r	ξ_r	M_r	N_r
1	1	$\{1\}$	$\{1,7\}$
2	1	$\{3\}$	$\{3,7\}$
3	1	$\{5\}$	$\{5,7\}$
4	2	$\{2\}$	$\{2,7\}$
5	2	$\{6\}$	$\{6,7\}$
6	3	$\{2\}$	$\{2,7\}$
7	4	$\{4\}$	$\{4,7\}$
8	5	$\{1\}$	$\{1,7\}$
9	5	$\{2\}$	$\{2,7\}$
10	5	$\{6\}$	$\{6,7\}$
11	6	$\{1\}$	$\{1,7\}$
12	6	$\{3\}$	$\{3,7\}$
13	7	$\{1\}$	$\{1,7\}$
14	7	$\{4\}$	$\{4,7\}$
15	7	$\{5\}$	$\{5,7\}$
16	8	$\{2\}$	$\{2,7\}$
17	8	$\{5\}$	$\{5,7\}$

Tasks		Processors (Π)								"Demand"
		p=1	p=2	p=3	p=4	p=5	p=6	p=7	p=8	
	t=1	1				1	1	1		1
	t=2		1	1		1			1	1
	t=3	1					1			1
Θ	t=4				1			1		1
	t=5	1						1	1	1
	t=6		1			1				1
$\{\theta+1\}$	t=7	1	1	1	1	1	1	1	1	11
"Supply"		3	2	1	1	3	2	3	2	17

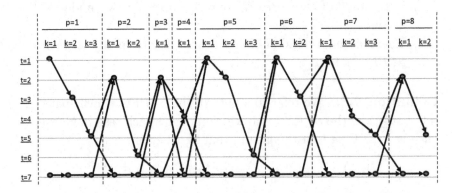

Figure 7.3. Numerical illustration of the structure of the *SPPFG*.

Remark 7.2. *It follows directly from definitions that*:

(1) There exists a one-to-one correspondence between *SPP paths* of the *SPPFG* and feasible SPP solutions.

(2) There exists a one-to-one correspondence between *SPP paths* of the *SPPFG* and feasible solutions to *Problem SPP*.

(3) There exists a one-to-one correspondence between *SPP paths* of the *SPPFG* and feasible solutions to *Problem TP*.

SPP paths are illustrated in Figure 7.4 for the numerical example shown in Figures 7.1 and 7.3. The stage-spanning path shown in Figure 7.4.a is a *SPP path*, and corresponds to the SPP solution in which the processors used are Processors 1, 2, and 4. The partial (i.e., non-spanning with respect to the set of *stages*) path shown in

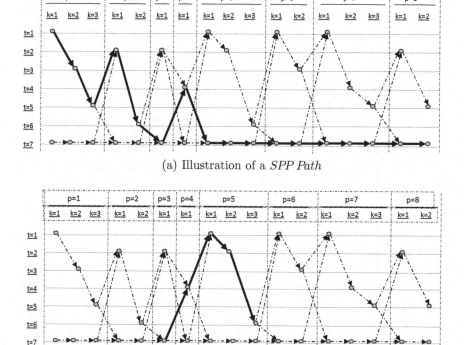

(a) Illustration of a *SPP Path*

(b) Illustration of infeasibility for a *SPP Path*

Figure 7.4. Numerical illustration of *SPP paths*.

Figure 7.4.b corresponds to SPP solutions in which both Processors 4 and 5 are used. It is easy to verify that there exists no *SPP path* in the graph which comprises this partial path, which is consistent with the fact that there exists no feasible SPP solution in which both Processors 4 and 5 are used.

Theorem 7.2. *A given SPP path of the SPPFG cannot be represented in terms of a convex combination of other SPP paths of the SPPFG.*

2.4. *LP model of the SPP*

Assumption 7.2. *We assume without loss of generality that the number of stages of the SPPFG is greater than 5 (i.e., $n \geq 6$).*

The variables and constraints of our proposed LP model for the SPP is essentially as shown in Chapter 2 of this book for the TSP. Note that because of the redefinition of the flow graph, the variables and constraints in this section model our "flows" over the SPPFG.

The costs we attach to arc $[i, r, j]$ of the SPPFG are as shown in (7.10), and are illustrated in Figure 7.5.

$$
d_{[i,r,j]} := \begin{cases}
c_{\xi_r} & \text{if } \xi_r < \gamma - 1; \ r = f_{\xi_r}; \ i \neq \theta + 1; \\
c_{\xi_r} & \text{if } \xi_r = \gamma - 1; \ r = f_{\xi_r} < \ell_{\xi_r}; \ i \neq \theta + 1; \\
c_{\xi_r} + c_{\xi_{r+1}} & \text{if } \xi_r = \gamma - 1; \ r = f_{\xi_r} = \ell_{\xi_r}; \ i, j \neq \theta + 1; \\
c_{\xi_r} & \text{if } \xi_r = \gamma - 1; \ r = f_{\xi_r} = \ell_{\xi_r}; \ i \neq \theta + 1; \ j = \theta + 1; \\
c_{\xi_{r+1}} & \text{if } \xi_r = \gamma - 1; \ r = \ell_{\xi_r} > f_{\xi_r}; \ i = \theta + 1; \ j \neq \theta + 1; \\
0 & \text{otherwise.}
\end{cases}
$$

$$(7.10)$$

In our overall LP model, we associate "costs" of 0 to the y-variables, respectively. The costs associated with the z-variables of our LP model are as follows:

$$\forall (p, r, s) \in R^3 : p < r < s,$$

$$\forall (u, v, i, j, k, t) \in (\Omega, F_p(u), \Omega, F_r(i), \Omega, F_s(k)),$$

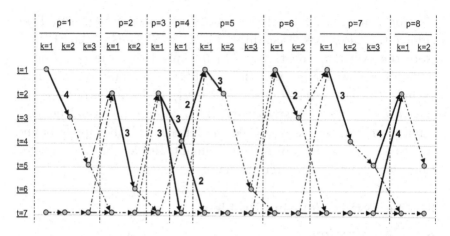

Figure 7.5. Illustration of the costs associated with arcs of the SPPFG.

$$e_{[u,p,v][i,r,j][k,s,t]} := \begin{cases} d_{[u,p,v]} + d_{[i,r,j]} + d_{[k,s,t]} & \text{if } p = 1;\ r = 2;\ s = 3; \\ d_{[k,s,t]} & \text{if } p = 1;\ r = 2;\ s > 3; \\ 0 & \text{otherwise.} \end{cases}$$

The complete statement of our LP model for the SPP is as follows.

Problem 7.3 (Problem SPPLP).

Minimize:

$$\sum_{([u,p,v][i,r,j][k,s,t]) \in A^3 : p < r < s} e_{[u,p,v][i,r,j][k,s,t]} z_{[u,p,v][i,r,j][k,s,t]}$$

Subject to:

Constraints (2.1)–(2.9);

$$y_{[i,r,j][k,s,t]} = 0;\quad \forall\, ([i,r,j],[k,s,t]) \in A^2 :$$
$$((i \neq \theta + 1)\ \text{and}\ (k = i))\ \text{or}$$
$$((i \neq \theta + 1)\ \text{and}\ (t = i))\ \text{or}$$
$$((s \neq r + 1)\ \text{and}\ (j \neq \theta + 1)\ \text{and}\ (t = j))\ \text{or}$$
$$((j \neq \theta + 1)\ \text{and}\ (k = j));\qquad (7.12)$$

$$y_{[i,r,j][k,s,t]} \geq 0,\quad [i,r,j],[k,s,t] \in A;\qquad (7.13)$$

$$z_{[u,p,v][i,r,j][k,s,t]} \geq 0,\quad [u,p,v],[i,r,j],[k,s,t] \in A.\qquad (7.14)$$

Constraints (7.12) are the "analogs" of constraints (2.10). They enforce the "no re-visit" condition for the *SPP paths* of the SPPFG. Note that whereas constraints (2.10) apply to all the *levels* of its underlying flow graph (i.e., the TSPFG), constraints (7.12) must exclude the last level (the *dummy level*) of its underlying graph (i.e., the SPPFG). Constraints (7.13)–(7.14) are the usual nonnegativity constraints. As discussed in Remark 3.1, the upper bound constraints on the variables are redundant. Hence, they are omitted from the LP problem statement here.

3. The Vertex Coloring Problem

3.1. *Introduction*

The Vertex Coloring Problem (VCP) can be described as follows. Let $\mathcal{G}(\mathcal{V}, \mathcal{E})$ be an undirected graph, with vertex (node) set $\mathcal{N}$, and edge set $\mathcal{E}$. A subset $\mathcal{N}' \subseteq \mathcal{N}$ of vertices of $\mathcal{G}$ is called an *independent set* if no two members of $\mathcal{N}'$ are adjacent. For a given integer $k>0$, a *k-coloring of* $\mathcal{G}$ is a partitioning of $\mathcal{N}$ into k *independent sets*. VCP seeks to determine the smallest integer $\chi(\mathcal{G}) > 0$, known as the *chromatic number* of $\mathcal{G}$, such that $\mathcal{G}$ admits a $\chi(\mathcal{G})$-*coloring*.

The history of the VCP goes back to the 19th century, when it was first posed as a mathematical problem in the context of coloring geographical maps (see Fritsch and Fritsch (1998)). However, the problem has applications in many other contexts. Examples of some of these in the recent literature include contexts of cloth/fabric design (Govindaraju *et al.* (2005)), communication network design (Oliveira *et al.* (2005); Woo *et al.* (2002)), computation of derivative matrices (Gebremedhin *et al.* (2005)), crew scheduling (Gamache *et al.* (2007)), image segmentation (Gomez *et al.* (2007)), logic circuits design (Kania and Kulisz (2007)), radio frequency identification systems design (Saygin *et al.* (2006)), robotics planning and scheduling (Demange *et al.* (2009)), satellite range scheduling (Zufferey *et al.* (2008)), sequencing of stamping operations (Chu *et al.* (2008)), spectrum assignments in wireless communication systems (Peng *et al.* (2006)), timetabling (Burke *et al.* (2007); de Werra (1985); Dowsland and Thompson (2005)), and wavelength assignments in optical networks (Noronha and Ribeiro (2006)).

The NP-Completeness of the VCP has been known since the 1970's (Karp (1972)). Hence, the focus of research aimed at developing solution procedures has been heuristics and enumerative procedures. Good reviews of these can be found in Pardalos *et al.* (1999), Laguna and Marti (2001), Galinier and Hertz (2006), and Malaguti *et al.* (2008). Some closely-related variants are also reviewed in Calamoneri (2006). In this section, we apply the results of the previous chapters of this book to present an LP formulation of the VCP. The complexity orders of the number of variables and the number of constraints of the proposed LP are $O\left(\nu^6 \cdot \varsigma^2\right)$ and $O\left(\nu^5 \cdot \varsigma^2\right)$, respectively, where ν and ς are the number of nodes and the number of available colors in the VCP instance, respectively.

We will first give an overview of the standard IP formulation of the VCP in Section 3.2. Then, we will propose a bipartite network flow (BNF)-based reformulation of the problem and illustrate it with a numerical example in Section 3.3. Our path-based formulation of the problem will be discussed in Section 3.4. Finally, the proposed LP model for the VCP will be discussed in Section 3.5.

3.2. Traditional IP model of the VCP

Definition 7.4. We refer to a set of color assignments that satisfies the VCP constraints, as a "feasible coloring of (the VCP graph)" $\mathcal{G}$.

Notation 7.3 (VCP parameters).

(1) ν: Number of vertices of *Graph* $\mathcal{G}$;
(2) ς: Number of available colors/labels;
(3) $\mathcal{V} := \{v_1, v_2, \ldots, v_\nu\}$ (Set of vertices of *Graph* $\mathcal{G}$);
(4) $\mathcal{N} := \{1, 2, \ldots, \nu\}$ (Set of indices for the vertices of *Graph* $\mathcal{G}$);
(5) $\mathcal{E} := \{(k, t) \in \mathcal{N}^2 : v_k$ and v_t are adjacent to each other$\}$ (Set of edges of *Graph* $\mathcal{G}$);
(6) $L := \{c_1, c_2, \ldots, c_\varsigma\}$ (Set of available colors/labels);
(7) $\mathcal{C} := \{1, \ldots, \varsigma\}$ (Index set for the colors);
(8) $\overline{\mathcal{E}} := \{(k, t) \in \mathcal{N}^2 \setminus \mathcal{E} : k \neq t\}$ (Set of edges of the complementary graph, $\overline{\mathcal{G}}(\mathcal{N}, \overline{\mathcal{E}})$, of $\mathcal{G}(\mathcal{N}, \mathcal{E})$).

Let u_j $(j \in \mathcal{C})$ denote a binary 0/1 variable that is equal to 1 iff c_j is used. Let o_{ij} $((i, j) \in (\mathcal{N}, \mathcal{C}))$ denote a binary 0/1 variable that is equal to 1 iff vertex v_i receives color c_j. The standard IP formulation of the VCP is as follows (see Coll *et al.* (2002); or Kochenberger *et al.* (2005), for example):

Problem 7.4 (*Problem VCP*).

minimize:

$$\zeta(u,o) := \sum_{j \in C} u_j, \tag{7.15}$$

subject to:

$$\sum_{j \in C} o_{ij} = 1; \quad i \in \mathcal{N} \tag{7.16}$$

$$o_{kj} + o_{tj} \leq u_j; \quad (k,t) \in \mathcal{E}, \quad j \in C \tag{7.17}$$

$$u_j, \ o_{ij} \in \{0,1\}; \quad j \in C, \quad i \in \mathcal{N}. \tag{7.18}$$

The objective function of *Problem VCP* (i.e., (7.15)) minimizes the number of *colors* used. Constraints (7.16) ensure that each node of the VCP graph, $\mathcal{G}$, receives exactly one color. Constraints (7.17) enforce the condition that adjacent nodes of $\mathcal{G}$ cannot receive the same *color*. Constraints (7.18) are the usual binary requirements on the variables. Clearly, there exists a one-to-one correspondence between the points of *Problem VCP* and the feasible colorings of the VCP graph, $\mathcal{G}$.

Definition 7.5. Let

$$P_0 := \left\{ (u,o) \in \mathcal{R}^{\varsigma(\nu+1)} : (u,o) \text{ satisfies } (7.16)\text{--}(7.18) \right\}. \tag{7.19}$$

We refer to $Conv(P_0)$ as the "VCP polytope".

We will now discuss our network flow-based reformulation of *Problem VCP*.

3.3. BNF-based model of the VCP

Assumption 7.3. *We assume, without loss of generality, that:*

(1) *Graph $\mathcal{G}$ has been augmented with a "dummy" node, denoted $v_{\nu+1}$;*
(2) *Dummy node $v_{\nu+1}$ is not adjacent to any node in $\mathcal{N}$;*
(3) *Dummy node $v_{\nu+1}$ can be "colored" multiple times, using one or more of the available colors;*
(4) *The nodes (including the dummy node $v_{\nu+1}$) which receive a given color $c_j \in L$, do so in order, with a lower-indexed node receiving the color before a higher-indexed node.*

Notation 7.4.

(1) $\forall i \in \mathcal{N} \cup \{\nu + 1\}$, $\mathcal{D}_i := \{j \in \mathcal{N} : (i, j) \notin \mathcal{E}\}$ (Set of indices of the nodes that are non-adjacent to node i in the augmented VCP graph);
(2) $\forall j \in \mathcal{C}$, $K_j := \{1, \ldots, \nu\}$ (Index set for the order in which color $c_j \in L$ is used);
(3) $\forall j \in \mathcal{C}$, $\overline{u}_j$ denotes a binary $0/1$ variable that is equal to 1 iff c_j is **not** used; (i.e., $u_j + \overline{u}_j = 1$);
(4) $\forall \langle i \in \mathcal{N} \cup \{\nu + 1\}; \ j \in \mathcal{C}; \ k \in K_j : K \leq i \rangle$, x_{ijk} denotes a non-negative variable that is greater than zero iff v_i is the kth among the vertices which receive c_j.

Our BNF formulation of the VCP is as follows.

Problem 7.5 (*Problem VCPTP*).

maximize:

$$\zeta(\overline{u}, x) := \sum_{j \in \mathcal{C}} \overline{u}_j, \tag{7.20}$$

subject to:

$$\sum_{j \in \mathcal{C}} \sum_{k \in K_j : k \leq i} x_{ijk} = 1; \quad i \in \mathcal{N} \tag{7.21}$$

$$\sum_{j \in \mathcal{C}} \sum_{k \in K_j} x_{\nu+1,jk} = \nu(\varsigma - 1); \qquad\qquad (7.22)$$

$$\sum_{i \in (\mathcal{N} \cup \{\nu+1\})} x_{ijk} = 1; \quad j \in \mathcal{C}, \quad k \in K_j, \qquad (7.23)$$

$$\sum_{k \in K_j} x_{pjk} + \sum_{k \in K_j} x_{qjk} + \overline{u}_j \le 1; \quad (p,q) \in \mathcal{E}, \quad j \in \mathcal{C},$$

$$(7.24)$$

$$\overline{u}_j, \ x_{ijk} \in \{0,1\}; \quad j \in \mathcal{C}, \quad k \in K_j, \quad i \in \mathcal{N}, \qquad (7.25)$$

$$x_{\nu+1,jk} \ge 0; \quad j \in \mathcal{C}, \quad k \in K_j. \qquad\qquad (7.26)$$

The objective function (7.20) seeks to maximize the number of colors that are **not** used. Constraints (7.21) ensure (in light of constraints (7.25)) that each node of $\mathcal{N}$ is colored exactly once. Constraints (7.24) ensure that no node receives a color that is not used, and that no two adjacent nodes receive a same given color when that color is used. Constraints (7.22) and (7.23) properly account the number of times the colors are not used (by being assigned to the dummy node).

Problem VCPTP is illustrated in Example 7.1.

Example 7.1. *For $\varsigma = 3$ (i.e., three colors) and the VCP graph shown below (taken from Kochenberger et al. (2005)).*

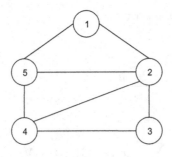

the BNF tableau form of the BNF-based formulation is:

$k =$	c_1					c_2					c_3					"Demand"
	1	2	3	4	5	1	2	3	4	5	1	2	3	4	5	
v_1	1					1					1					1
v_2	1	1				1	1				1	1				1
v_3	1	1	1			1	1	1			1	1	1			1
v_4	1	1	1	1		1	1	1	1		1	1	1	1		1
v_5	1	1	1	1	1	1	1	1	1	1	1	1	1	1	1	1
v_6	1	1	1	1	1	1	1	1	1	1	1	1	1	1	1	10
"Supply"	1	1	1	1	1	1	1	1	1	1	1	1	1	1	1	–

Each cell of the this tableau corresponds to an x-variable of Problem VCPTP. Specifically, variable x_{ijk} corresponds to the cell formed by row v_i and the kth column under c_j.

Definition 7.6. Let

$$P_1 := \left\{ (\overline{u}, x) \in \mathcal{R}^{\varsigma(\nu^2 + \nu + 1)} : (\overline{u}, x) \text{ satisfies } (7.21)\text{–}(7.26) \right\}.$$

We refer to $\mathrm{Conv}(P_1)$ as the "VCP Bipartite Network Flow (VCP BNF) polytope."

Theorem 7.3. *The following statements are true:*

(1) There exists a one-to-one correspondence between the points of the *VCP BNF polytope* and the points of the *VCP polytope*.

(2) There exists a one-to-one correspondence between the extreme points of the *VCP BNF polytope* and *feasible colorings* of the VCP graph, $\mathcal{G}$.

3.4. *Path-based reformulation of Problem VCPTP*

The graph which serves as the framework for our reformulation the *VCP BNF polytope*, $Conv(P_1)$, is illustrated in Example 7.2. We refer to this graph as the "VCP Flow Graph (VCPFG)". In the VCPFG, nodes consist of triplets $(i, j, k) \in ((\mathcal{N} \cup \{\nu + 1\}), \mathcal{C}, K_j)$. The arcs of the graph are specified through the explicit statements of the forward and backward stars of the nodes of the VCPFG, respectively. They correspond essentially to edges of the complementary graph $\overline{\mathcal{G}}(\mathcal{V},\overline{\mathcal{E}})$, of the VCP graph $\mathcal{G}(\mathcal{V},\mathcal{E})$.

Definition 7.7.

(1) The set of nodes of the VCPFG which correspond to a given pair $(j,k) \in (\mathcal{N}, K_j)$ is referred to as a *stage* of the graph.
(2) The set of nodes of the VCPFG which correspond to a given node $i \in (\mathcal{N} \cup \{\nu + 1\})$ is referred to as a *level* of the graph.

Notation 7.5 (VCPFG Notations).

(1) $M := \mathcal{N} \cup \{\nu + 1\}$;
(2) $n := \nu \times \varsigma$ (Number of stages of the VCPFG);
(3) $S := \{1, \ldots, n\}$ (Set of stages of the VCPFG);
(4) $R := S\backslash\{n\}$ (Set of stages of the VCPFG from which arcs originate);
(5) $\forall j \in \mathcal{C}, b_j := 1 + (j-1)\varsigma$ ("Stage index" of the pair $(j,1)$; i.e., index of the first stage which has color j);
(6) $\forall j \in \mathcal{C}, e_j := j\varsigma$ ("Stage index" of the pair (j,ν)); i.e., index of the last stage which has color j)
(7) $\forall r \in S, \mathcal{K}_r := \max\{j \in \mathcal{C}: b_j \le r\}$ ("Color" of stage r; i.e., index of the color associated with stage r);
(8) $\forall r \in S, N_r := \{[i,r] : i \in M; i > r - b_j\}$ (Set of nodes/vertices at stage r of the VCPFG);
(9) $\overline{N} : \cup_{r \in S}(N_r)$
(10) $\forall r \in S; \forall i \in M,$

$$F_r(i) := \begin{cases} M & \text{if } e_{\mathcal{K}_r} = r < n; \\ \{j : j \in N_r \cap \mathcal{D}_i\} & \text{if } b_{\mathcal{K}_r} \le r < e_{\mathcal{K}_r}; \\ r < n; \quad i \ne \nu + 1; \\ \{\theta + 1\} & \text{if } b_{\mathcal{K}_r} \le r < e_{\mathcal{K}_r}; r < n; \quad i = \nu + 1; \\ \varnothing & \text{for } r = n \end{cases}$$

(Forward star of node $[i,r]$ of the VCPFG;

(11) $\forall r \in S; \ \forall i \in \overline{N},$

$$B_r(i) := \begin{cases} \varnothing & \text{if } r = 1, \\ \{j : j \in F_{r-1}(i)\} & \text{if } r > 1 \end{cases}$$

(Backward star of node $[i, r]$ of the VCPFG);

(12) $A := \{[i, r, j] \in (M, R, M) : j \in F_r(i)\}$ (Set of arcs of the VCPFG).

The notation for the multipartite graph representation is illustrated in Example 7.2 for the VCP instance of Example 7.1.

Example 7.2. *For the VCP of Example* 7.1, *we have the following:*

• VCP Graph:

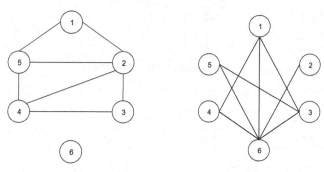

Augmented VCP graph   Complement of the augmented VCP graph

- "Colors" of the stages:

Stage, r	"Color", $\mathcal{K}_r$
$r \in \{1, 2, 3, 4, 5\}$	1
$r \in \{6, 7, 8, 9, 10\}$	2
$r \in \{11, 12, 13, 14, 15\}$	3

- Forward stars of the nodes of the VCPFG:

$Level, i$	Stage, r			
	$r \in \{1, 6, 11\}$	$r \in \{2, 7, 12\}$	$r \in \{3, 8, 13\}$	$r \in \{4, 9, 14\}$
$i = 1$	$\{3, 4, 6\}$	N/A	N/A	N/A
$i = 2$	$\{6\}$	$\{6\}$	N/A	N/A
$i = 3$	$\{5, 6\}$	$\{5, 6\}$	$\{5, 6\}$	N/A
$i = 4$	$\{6\}$	$\{6\}$	$\{6\}$	N/A
$i = 5$	$\{6\}$	$\{6\}$	$\{6\}$	$\{6\}$
$i = 6$	$\{6\}$	$\{6\}$	$\{6\}$	$\{6\}$

$Level, i$	Stage, r	
	$r \in \{5, 10\}$	$r = 15$
$i = 1$	N/A	$\varnothing$
$i = 2$	N/A	$\varnothing$
$i = 3$	N/A	$\varnothing$
$i = 4$	N/A	$\varnothing$
$i = 5$	$\{1, 2, 3, 4, 5, 6\}$	$\varnothing$
$i = 6$	$\{1, 2, 3, 4, 5, 6\}$	$\varnothing$

- Backward stars of the nodes of of the VCPFG:

$Level, i$	Stage, r		
	$r = 1$	$r \in \{6, 11\}$	$r \in \{2, 7, 12\}$
$i = 1$	$\varnothing$	$\{5, 6\}$	N/A
$i = 2$	$\varnothing$	$\{5, 6\}$	$\varnothing$
$i = 3$	$\varnothing$	$\{5, 6\}$	$\{1\}$
$i = 4$	$\varnothing$	$\{5, 6\}$	$\{1\}$
$i = 5$	$\varnothing$	$\{5, 6\}$	$\{3\}$
$i = 6$	$\varnothing$	$\{5, 6\}$	$\{1, 2, 3, 4, 5, 6\}$

	Stage, r		
Level, i	$r \in \{3, 8, 13\}$	$r \in \{4, 9, 14\}$	$r \in \{5, 10, 15\}$
$i = 1$	N/A	N/A	N/A
$i = 2$	N/A	N/A	N/A
$i = 3$	$\varnothing$	N/A	N/A
$i = 4$	$\varnothing$	$\varnothing$	N/A
$i = 5$	$\{3\}$	$\{3\}$	$\varnothing$
$i = 6$	$\{3, 4, 5, 6\}$	$\{3, 4, 5, 6\}$	$\{4, 5, 6\}$

Definition 7.8. We refer to a path of the VCPFG which spans the set of *stages* of the graph (i.e., a path of length $n - 1$), includes each node in $\mathcal{N}$ exactly once, and is such that the stages of pairs of nodes pertaining to adjacent vertices of $\mathcal{G}$ have different *colors*, as a *VCP path* of the graph. In other words, a set of arcs, $((i_1, 1, i_2), (i_2, 2, i_3), \ldots, (i_{n-1}, n - 1, i_n)) \in A^{n-1}$, is a *VCP path* of the VCPFG iff the following three conditions are satisfied:

(1) $\forall t \in \mathcal{N}, \exists p \in S : i_p = t$;
(2) $\forall (p, q) \in (S, S \backslash \{p\}) : (i_p, i_q) \in \mathcal{N}^2, i_p \neq i_q$;
(3) $\forall (p, q) \in S^2 : (i_p, i_q) \in \mathcal{E}, \mathcal{K}_p \neq \mathcal{K}_q$.

VCP paths are illustrated in Figures 7.6 and 7.7 for the VCP instance of Example 7.1. The spanning path shown in Figure 7.6 is a *VCP path* and corresponds to the *coloring* where vertices v_1 and v_4 receive *color* c_1, vertex v_2 receives *color* c_2, and vertices v_3 and v_5 receive *color* c_3. The partial (non-spanning) path shown in Figure 7.7 corresponds to a *coloring* in which *color* c_2 is applied to vertex v_2, and *color* c_1 is exclusively applied to vertex v_3. It is easy to verify that there exists no *VCP path* in the graph shown on the figure (Figure 7.7) that comprises this partial path (since either a *level* (other than the *dummy level*) is "revisited", or at least one *level* (other than the *dummy level*) cannot be "visited"),

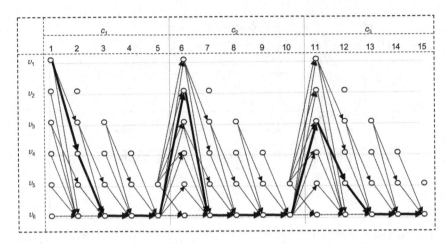

Figure 7.6. Illustration of a *VCP path* of the VCPFG.

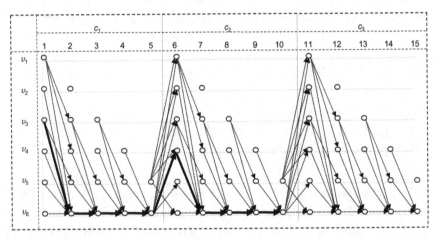

Figure 7.7. Illustration of infeasibility for a *VCP path* of the VCPFG.

which is consistent with the fact that there exists no *feasible coloring* of the problem instance comprising this partial assignment of *colors*.

Remark 7.3. *It follows directly from definitions that:*

(1) There exists a one-to-one correspondence between *VCP paths* of the VCPFG and *feasible colorings* of $\mathcal{G}$.

(2) There exists a one-to-one correspondence between the *VCP paths* of the VCPFG and the extreme points of the *VCP polytope*, $Conv(P_0)$.

(3) There exists a one-to-one correspondence between the *VCP paths* of the VCPFG and the extreme points of the *BNF polytope*, $Conv(P_1)$.

3.5. *LP model of the VCP*

Assumption 7.4. *We assume without loss of generality that number of colors, and the number of nodes of $\mathcal{G}$ are greater than or equal to 3, respectively (i.e., $\varsigma \geq 3$, and $\nu \geq 3$).*

3.5.1. *Objective function "costs"*

We use the variables as defined in Notation 2.2 of Chapter 2 of this book, but where the arcs and nodes are specified over the graph defined in Notation 7.5, in formulating our proposed model for the VCP. Since the VCP is essentially a feasibility problem (no costs are involved), one may seek to minimize the number of colors used (as in the standard VCP formulation, i.e., *Problem VCP*), or to maximize the number of colors not used (as in *Problem BNF*). These two alternatives are illustrated in Figures 7.8 and 7.9 respectively, for the illustrative problem of Example 7.1.

We use the maximization form illustrated in Figure 7.9, in stating our overall LP model for the VCP. In doing this, to each y-variable we associate a "cost" of 0, and to the z-variables we associate the "costs" defined below.

Notation 7.6 (VCPLP "costs"). *The "costs" associated with the z-variables are as follows:*

$$\forall ([u,p,v],[i,r,j],[k,s,t]) \in A^3 : p < r < s,$$

$$\delta_{[u,p,v][i,r,j][k,s,t]} := \begin{cases} 1 & \text{if } p = b_{\xi_r}; \ r = p+1; \ s = p+2; \ u = \nu+1 \\ 0 & \text{otherwise.} \end{cases}$$

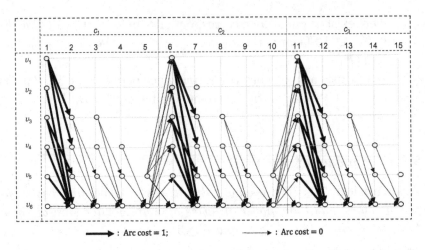

Figure 7.8. Illustration of arc costs for an objective function to be minimized.

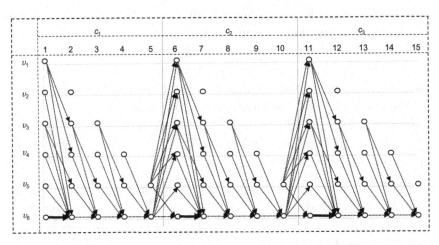

Figure 7.9. Illustration of arc costs for an objective function to be maximized.

3.5.2. *Statement of the VCP LP model*

Problem 7.6 (Problem VCPLP).

Maximize:

$$\sum_{([u,p,v][i,r,j][k,s,t])\in A^3:p<r<s} \delta_{[u,p,v][i,r,j][k,s,t]} z_{[u,p,v][i,r,j][k,s,t]} \quad (7.27)$$

Subject to:

Constraints (2.1)–(2.9);

$$y_{[i,r,j][k,s,t]} = 0; \quad \forall([i,r,j],[k,s,t]) \in A^2 :$$

$$(((i,k) \in \mathcal{E}) \quad and \quad (\xi_r = \xi_s)) \quad or$$

$$(((i,t) \in \mathcal{E}) \quad and \quad (\xi_r = \xi_{s+1})) \quad or$$

$$(((j,k) \in \mathcal{E}) \quad and \quad (\xi_{r+1} = \xi_s)) \quad or$$

$$(((j,t) \in \mathcal{E}) \quad and \quad (\xi_{r+1} = \xi_{s+1})); \quad (7.28)$$

$$y_{[i,r,j][k,s,t]} = 0; \quad \forall([i,r,j],[k,s,t]) \in A^2 :$$

$$((i \neq \nu+1) \ and \ (t=i)) \quad or$$

$$((i \neq \nu+1) \ and \ (t=i)) \quad or$$

$$((s \neq r+1) \ and \ (j \neq \nu+1) \ and \ (k=j)) \quad or$$

$$((j \neq \nu+1) \ and \ (k=j)); \quad (7.29)$$

$$y_{[i,r,j][k,s,t]} \geq 0, \ [i,r,j],[k,s,t] \in A; \quad (7.30)$$

$$z_{[u,p,v][i,r,j][k,s,t]} \geq 0 \ [u,p,v],[i,r,j],[k,s,t] \in A. \quad (7.31)$$

□

Constraints (7.29) are similar to constraints (7.12) for the SPP LP model. They are the "analogs" of constraints (2.10) for the TSP LP model. They enforce the "no revisit" condition for all the *levels* of the VCPFG, except for the *dummy level*. Constraints (7.28) have no "analogs" for either the TSP or the SPP LP models. Note that ξ_r in these constraints indicates the "color" of (associated with) stage r of the VCPFG. Hence, these constraints restrict each of the variables which would involve an assignment of a same color to adjacent nodes of the *VCP Graph* $\mathcal{G}$ to zero. Hence, constraints (7.28) essentially exclude variables which would involve an assignment of a same color to adjacent nodes of the *VCP Graph* $\mathcal{G}$ from the LP model. Constraints (7.30)–(7.31) are the usual non-negativity constraints. As in the TSP and VCP models, the upper bound constraints on the variables are omitted from the LP statement, since those constraints are redundant.

4. The Multiple Traveling Salesman Problem (mTSP)

4.1. *Introduction*

A generalization of the TSP (see Chapter 1) which has a very wide-range of applicability is the mTSP. In the mTSP, each of c cities must be visited by exactly one of s ($s > 1$) salesmen. All of the salesmen can be based at a single depot (*single depot* mTSP), or they may be "homed" at different depots (*multi-depot* mTSP; denoted MmTSP). If each salesman is required to return to his/her "home" base/depot at the end of his/her travels, the problem is referred to as the *fixed destination* MmTSP. Otherwise, the problem is referred to as the *nonfixed destination* MmTSP. Also, if there is no requirement that every salesman be activated, then fixed costs are typically associated with the salesmen and included in the cost-minimization objective of the problem. Good discussions of these and other variations of the problem can be found in Bektas (2006) and Kara and Bektas (2006).

Contexts in which the mTSP has been applied are numerous. Bektas (2006) for example, discusses applications in contexts of combat mission planning, transportation planning, print scheduling, satellite surveying system design, and workforce planning. Beyond specific applications however, one could easily argue that most of the practical contexts in which the TSP has been applied could be more realistically modeled as mTSP's. Hence, the range of applicability of the problem is indeed very broad.

The *NP-Hardness* (see Garey and Johnson (1969)) of the mTSP follows directly from the fact it is a generalization of the TSP. Hence, solution methods have been mostly heuristic procedures (see Bektas (2006) and Ghafurian and Javadian (2011)). We classify these into two broad groups which we label as the "transformation-based" and the "direct" heuristics groups, respectively. The idea of the "transformation-based" heuristics is to transform the problem into a standard TSP on expanded graphs, and then to use TSP heuristics to solve it (see Betkas (2006)). The "direct" heuristics tackle the problem in its natural form. They include evolutionary, genetic, *k-opt*, neural network, simulated annealing, and tabu search procedures, respectively (see Bektas (2006), and Ghafurian and Javadian (2011) for more detailed discussions). As far as we know, the exact procedures which have been developed are the cutting planes approach of Laporte and Norbert (1980), and the branch-and-bound approaches of Ali and Kennington (1986), Gavish and Srikanth (1986), and Gromicho *et al.* (1992) (see Bektas (2006)).

A general limitation of the existing literature is the fragmentation of models over the different types of mTSP's discussed above. In general, models developed for one type of mTSP cannot be applied in a straightforward manner to other types. Also, to the best of our knowledge, except for the VRP model of Christofides *et al.* (1981), and the *fixed destination* MmTSP IP model of Kara and Bektas (2006), none of the existing models can be extended in a straightforward manner to handle differentiated travel costs for the salesmen. Differentiated travel costs are more realistic in many practical situations however, such as in contexts of routing/scheduling vehicles for example, where there may be differing pay rates for drivers, vehicle types, and/or transportation modes.

In this section, we will present a LP model for the general form of the mTSP in which the intersite travel costs associated with the salesmen are differentiated, there are multiple depots from which travels start (i.e., the problem considered is the MmTSP), salesmen are required to return to their respective "home" bases at the end of their travels (i.e., destinations are *fixed*), and the number of salesmen to be activated is a decision variable. The complexity orders of the number of variables and the number of constraints of the proposed LP are $O\left(\mathfrak{c}^9 \cdot \mathfrak{s}^3\right)$ and $O\left(\mathfrak{c}^8 \cdot \mathfrak{s}^3\right)$, respectively, where $\mathfrak{c}$ and $\mathfrak{s}$ are the numbers of customer sites and salesmen in the MmTSP instance, respectively. In formulating our proposed LP, we first develop a BNF-based model of the problem. Then, we use the path-based modeling framework described in the previous chapters of this book and the previous sections of this chapter. The approach is illustrated with a numerical example.

Definition 7.9 (*"MmTSP schedule"*). We will refer to a feasible solution to the fixed destination MmTSP as an "MmTSP schedule."

The notation we use for the mTSP is as follows.

Notation 7.7 (MmTSP notation).

(1) $\mathfrak{d}$: Number of depot sites/nodes;
(2) $\mathbb{D} := \{1, 2, \ldots, \mathfrak{d}\}$ (index set for the depot sites);
(3) $\mathfrak{c}$: Number of customer sites/nodes;
(4) $\mathbb{C} := \{1, 2, \ldots, \mathfrak{c}\}$ (index set for the customer sites);
(5) $\mathfrak{s}$: Number of salesmen;
(6) $\mathbb{S} := \{1, 2, \ldots, \mathfrak{s}\}$ (index set for the salesmen);
(7) $\forall p \in \mathbb{S}$, $\mathfrak{b}_p$: Index of the starting base (or "home" depot) for salesman p ($\mathfrak{b}_p \in \mathbb{D}$);
(8) $\forall p \in \mathbb{S}$, $\mathfrak{f}_p$: Fixed cost associated with the activation of salesman p;
(9) $\forall p \in \mathbb{S}$, $\forall (i, j) \in (\mathbb{D} \cup \mathbb{C})^2$, $\mathfrak{e}_{pij}$: Cost of travel from site i to site j by salesman p;
(10) A *MmTSP schedule* wherein salesman p visits m_p customers with $i_{p,k}$ being the kth customer visited will be denoted by the ordered set $\{(p, i_{p,k}) : p \in \overline{\mathbb{S}}, \, k = 1, \ldots, m_p\}$ where $\overline{\mathbb{S}} \subseteq \mathbb{S}$ denotes the subset of *activated* salesmen.

Assumption 7.5. *We assume, without loss of generality, that:*

(1) $\mathfrak{c} \geq 5$;
(2) $\mathfrak{d} \geq 1$;
(3) $\forall j \in \mathbb{D}$, $\{p \in \mathbb{S} : \mathfrak{b}_p = j\} \neq \varnothing$;
(4) $\forall p \in \mathbb{S}$, $\forall i \in \mathbb{C}$, $\mathfrak{e}_{pii} = \infty$;
(5) $\forall p \in \mathbb{S}$, $\forall (i, j) \in \mathbb{D}^2$, $\mathfrak{e}_{pij} = \infty$;
(6) The set of cutomers/customer sites has been augmented with a fictitious customer/site, indexed as $\overline{\mathfrak{c}} := \mathfrak{c} + 1$, with $\mathfrak{e}_{p,\overline{\mathfrak{c}},\overline{\mathfrak{c}}} = 0$

for all $p \in \mathbb{S}$, $\mathfrak{e}_{p,i,\bar{\mathfrak{c}}} = \mathfrak{e}_{p,i,\mathfrak{b}_p}$ for all $(p,i) \in (\mathbb{S}, \mathbb{C})$, and $\mathfrak{e}_{p,\bar{\mathfrak{c}},i} = \infty$ for all $(p,i) \in (\mathbb{S}, \mathbb{C})$;

(7) Fictitious customer site $\bar{\mathfrak{c}}$ can be visited multiple times by one or more of the traveling salesmen in any *MmTSP schedule*.

4.2. BNF-based model of MmTSP schedules

As far as we know, the BNF-based model we propose in this section is the first network flow-based model for the MmTSP. We will first present the model. Then, we will illustrate it with a numerical example.

Notation 7.8.

(1) $M := \mathbb{C} \cup \{\bar{\mathfrak{c}}\} = \mathbb{C} \cup \{\mathfrak{c}+1\}$;
(2) $\forall p \in \mathbb{S}$, $\mathbb{T}_p := \{1, \ldots, \mathfrak{c}\}$ (index set for the order (or "times") of visits for salesman p);
(3) $\forall p \in \mathbb{S}$, $\forall i \in M$, $\forall t \in \mathbb{T}_p$, $x_{p,i,t}$ denotes a nonnegative variable that is greater than zero iff i is the tth customer to be visited by salesman p.

Definition 7.10 ("MmTSP *BNF-based Polytope*"). Let

$$P_1 := \big\{ x \in \mathbb{R}^{\mathfrak{s}\mathfrak{c}\bar{\mathfrak{c}}} : x \text{ satisfies } (7.32)\text{–}(7.37) \big\},$$

where (7.32)–(7.37) are specified as follows:

$$\sum_{p \in \mathbb{S}} \sum_{t \in \mathbb{T}_p} x_{p,i,t} = 1; \quad i \in \mathbb{C}; \tag{7.32}$$

$$\sum_{p \in \mathbb{S}} \sum_{t \in \mathbb{T}_p} x_{p,\bar{\mathfrak{c}},t} = (\mathfrak{s}-1)\mathfrak{c}; \tag{7.33}$$

$$\sum_{i \in M} x_{p,i,t} = 1; \quad p \in \mathbb{S}, \, t \in \mathbb{T}_p, \tag{7.34}$$

$$x_{p,\bar{\mathfrak{c}},t-1} - x_{p,\bar{\mathfrak{c}},t} \leq 0; \quad p \in \mathbb{S}, \, t \in \mathbb{T}_p : t > 1, \tag{7.35}$$

$$x_{pit} \in \{0,1\}; \quad p \in \mathbb{S}, \, i \in \mathbb{C}, \, t \in \mathbb{T}, \tag{7.36}$$

$$x_{p,\bar{\mathfrak{c}},t} \geq 0; \quad p \in \mathbb{S}, \, t \in \mathbb{T}_p. \tag{7.37}$$

We refer to $Conv(P_1)$ as the "BNF-based Polytope".

Theorem 7.4. *There exists a one-to-one mapping of the points of P_1 (i.e., the extreme points of the BNF-based Polytope) onto the MmTSP schedules.*

The *BNF*-based formulation is illustrated in Example 7.3.

Example 7.3.

- *Fixed destination MmTSP with:*
  - $\mathfrak{d} = 2$, $\mathbb{D} = \{1, 2\}$;
  - $\mathfrak{s} = 2$, $\mathbb{S} = \{1, 2\}$, $\mathfrak{b}_1 = 1$, $\mathfrak{b}_2 = 2$;
  - $\mathfrak{c} = 5$, $\mathbb{C} = \{1, 2, 3, 4, 5\}$;

- *BNF tableau form of the BNF-based formulation (where entries in the body are "technical coefficients", and entries in the margins are "right-hand-side values"):*

time of visit, $t =$	*salesman "1"*					*salesman "2"*					*"Demand"*
	1	2	3	4	5	1	2	3	4	5	
customer "1"	1	1	1	1	1	1	1	1	1	1	1
customer "2"	1	1	1	1	1	1	1	1	1	1	1
customer "3"	1	1	1	1	1	1	1	1	1	1	1
customer "4"	1	1	1	1	1	1	1	1	1	1	1
customer "5"	1	1	1	1	1	1	1	1	1	1	1
customer "6"	1	1	1	1	1	1	1	1	1	1	5
"Supply"	1	1	1	1	1	1	1	1	1	1	–

- *Illustrations of Theorem 7.4:*

- *Illustration 1:*
Let the MmTSP schedule be: $\{(1,1), (1,3), (1,2), (2,5), (2,4)\}$.

The unique point of P_1 corresponding to this schedule is obtained by setting the entries of x as follows:

$$\forall (i, t) \in (M, \mathbb{T}_1), \ x_{1,i,t}$$
$$= \begin{cases} 1 & \text{if } (i, t) \in \{(1,1), (3,2), (2,3), \{6,4\}, (6,5)\} \\ 0 & \text{otherwise} \end{cases}$$

$\forall (i,t) \in (M, \mathbb{T}_2), \ x_{2,i,t}$

$$= \begin{cases} 1 & \text{if } (i,t) \in \{(5,1),(4,2),(6,3),\{6,4\},(6,5)\} \\ 0 & \text{otherwise} \end{cases}$$

This solution can be shown in tableau form as follows (where only non-zero entries of x are shown):

	salesman "1"					salesman "2"				
time of visit, $t =$	1	2	3	4	5	1	2	3	4	5
customer "1"	1									
customer "2"			1							
customer "3"		1								
customer "4"							1			
customer "5"						1				
customer "6"				1	1			1	1	1

- *Illustration 2:*
  Let $x \in P_1$ be as follows:

$\forall (i,t) \in (M, \mathbb{T}_1), \ x_{1,i,t}$

$$= \begin{cases} 1 & \text{for } (i,t) \in \{(6,1),(6,2),(6,3),\{6,4\},(6,5)\} \\ 0 & \text{otherwise} \end{cases}$$

$\forall (i,t) \in (M, \mathbb{T}_2), \ x_{2,i,t}$

$$= \begin{cases} 1 & \text{for } (i,t) \in \{(3,1),(5,2),(1,3),\{4,4\},(2,5)\} \\ 0 & \text{otherwise} \end{cases}$$

The unique MmTSP schedule corresponding to this point is
$\{(2,3),(2,5),(2,1),(2,4),(2,2)\}$.

4.3. *Path representation of BNF-based solutions*

In this section, we develop a path representation of the extreme points of the *BNF-based Polytope* (i.e., the points of P_1). The framework for this representation is the multipartite digraph illustrated in

Example 7.4. We refer to this graph as the "MmTSP Flow Graph (MmTSPFG)". The nodes of the MmTSPFG correspond to the variables of the *BNF*-based formulation (i.e., the "cells" of the *BNF-based* tableau). The arcs of the graph (roughly) represent the intersite movements at consecutive *times of travel*, respectively.

Definition 7.11.

(1) The set of nodes of the MmTSPFG which correspond to a given pair $(p, k) \in (\mathbb{S}, \mathbb{T}_p)$ is referred to as a *stage* of the graph.
(2) The set of nodes of the MmTSPFG which correspond to a given customer site $i \in M$ is referred to as a *level* of the graph.

In order to simplify the presentation, we perform a sequential re-indexing of the stages of the MmTSPFG and formalize the specifications of the nodes and arcs accordingly, as follows.

Notation 7.9 (MmTSPFG formalisms).

(1) $n := \mathfrak{sc}$ (Number of stages of the MmTSPFG);
(2) $S := \{1, \ldots, n\}$ (Set of stages of the MmTSPFG);
(3) $R := S \backslash \{n\}$ (Set of stages of the MmTSPFG with positive outdegree nodes);
(4) $\forall p \in \mathbb{S}$, $\underline{\mathfrak{r}}_p := ((p-1)\mathfrak{c} + 1)$ (Sequential re-indexing of stage $(p, 1)$);
(5) $\forall p \in \mathbb{S}$, $\bar{\mathfrak{r}}_p := p\mathfrak{c}$ (Sequential re-indexing of stage $(p, \mathfrak{c})$);
(6) $\forall r \in S$, $\mathfrak{p}_r := \max\{p \in \mathbb{S} \colon \underline{\mathfrak{r}}_p \leq r\}$ (Index of the salesman associated with stage r);
(7) $\overline{N} := \{(i, r) : i \in M, r \in S\}$ (Set of nodes/vertices of the MmTSPFG);
(8) $\forall r \in S$, $\forall i \in M$,

$$
F_r(i) := \begin{cases} M \backslash \{i\} & \text{for } r < n; \ i \in \mathbb{C}; \\ \{\bar{c}\} & \text{for } r < \bar{\mathfrak{r}}_{\mathfrak{p}_r}; \ i = \bar{c}, \\ M & \text{for } \bar{\mathfrak{r}}_{\mathfrak{p}_r} = r < n; \ i = \bar{c}, \\ \varnothing & \text{for } r = n \end{cases}
$$

(Forward star of node (i, r) of the MmTSPFG);

(9) $\forall r \in S$, $\forall i \in M$,

$$B_r(i) := \begin{cases} \varnothing & \text{for } r = 1 \\ \{j \in M : i \in F_{r-1}(j)\} & \text{for } r > 1 \end{cases}$$

(Backward star of node (i, r) of the MmTSPFG);

(10) $A := \{[i, r, j] \in (M, R, M) : j \in F_r(i)\}$ (Set of arcs of the MmTSPFG).

The notation for the multipartite graph representation is illustrated in Example 7.4 for the MmTSP instance of Example 7.3.

Example 7.4. *The multipartite graph representation of the MmTSP of Example 7.3 is summarized as follows*:

- $n = 2 \times 5 = 10$; $S = \{1, 2, \ldots, 10\}$; $R = \{1, \ldots, 9\}$;

- Stage indices for the salesmen:

Salesman, p	First stage, $\underline{\mathfrak{r}}_p$	Last stage, $\bar{\mathfrak{r}}_p$
1	1	5
2	6	10

- Salesman index for the stages:

Stage, r	Salesman index, $\mathfrak{p}_r$
$r \in \{1, 2, 3, 4, 5\}$	1
$r \in \{6, 7, 8, 9, 10\}$	2

- Forward stars of the nodes of the MmTSPFG:

Level, i	\multicolumn{10}{c}{Stage, r}									
	1	2	3	4	5	6	7	8	9	10
$i = 1$	$M \backslash \{1\}$	$M \backslash \{1\}$	$M \backslash \{1\}$	$M \backslash \{1\}$	$M \backslash \{1\}$	$M \backslash \{1\}$	$M \backslash \{1\}$	$M \backslash \{1\}$	$M \backslash \{1\}$	$\varnothing$
$i = 2$	$M \backslash \{2\}$	$M \backslash \{2\}$	$M \backslash \{2\}$	$M \backslash \{2\}$	$M \backslash \{2\}$	$M \backslash \{2\}$	$M \backslash \{2\}$	$M \backslash \{2\}$	$M \backslash \{2\}$	$\varnothing$
$i = 3$	$M \backslash \{3\}$	$M \backslash \{3\}$	$M \backslash \{3\}$	$M \backslash \{3\}$	$M \backslash \{3\}$	$M \backslash \{3\}$	$M \backslash \{3\}$	$M \backslash \{3\}$	$M \backslash \{3\}$	$\varnothing$
$i = 4$	$M \backslash \{4\}$	$M \backslash \{4\}$	$M \backslash \{4\}$	$M \backslash \{4\}$	$M \backslash \{4\}$	$M \backslash \{4\}$	$M \backslash \{4\}$	$M \backslash \{4\}$	$M \backslash \{4\}$	$\varnothing$
$i = 5$	$M \backslash \{5\}$	$M \backslash \{5\}$	$M \backslash \{5\}$	$M \backslash \{5\}$	$M \backslash \{5\}$	$M \backslash \{5\}$	$M \backslash \{5\}$	$M \backslash \{5\}$	$M \backslash \{5\}$	$\varnothing$
$i = 6$	$\{6\}$	$\{6\}$	$\{6\}$	$\{6\}$	M	$\{6\}$	$\{6\}$	$\{6\}$	$\{6\}$	$\varnothing$

- Backward stars of the nodes of the MmTSPFG:

Level, i	1	2	3	4	5	6	7	8	9	10
						Stage, r				
$i=1$	$\varnothing$	$\mathbb{C}\backslash\{1\}$	$\mathbb{C}\backslash\{1\}$	$\mathbb{C}\backslash\{1\}$	$\mathbb{C}\backslash\{1\}$	$M\backslash\{1\}$	$\mathbb{C}\backslash\{1\}$	$\mathbb{C}\backslash\{1\}$	$\mathbb{C}\backslash\{1\}$	$\mathbb{C}\backslash\{1\}$
$i=2$	$\varnothing$	$\mathbb{C}\backslash\{2\}$	$\mathbb{C}\backslash\{2\}$	$\mathbb{C}\backslash\{2\}$	$\mathbb{C}\backslash\{2\}$	$M\backslash\{2\}$	$\mathbb{C}\backslash\{2\}$	$\mathbb{C}\backslash\{2\}$	$\mathbb{C}\backslash\{2\}$	$\mathbb{C}\backslash\{2\}$
$i=3$	$\varnothing$	$\mathbb{C}\backslash\{3\}$	$\mathbb{C}\backslash\{3\}$	$\mathbb{C}\backslash\{3\}$	$\mathbb{C}\backslash\{3\}$	$M\backslash\{3\}$	$\mathbb{C}\backslash\{3\}$	$\mathbb{C}\backslash\{3\}$	$\mathbb{C}\backslash\{3\}$	$\mathbb{C}\backslash\{3\}$
$i=4$	$\varnothing$	$\mathbb{C}\backslash\{4\}$	$\mathbb{C}\backslash\{4\}$	$\mathbb{C}\backslash\{4\}$	$\mathbb{C}\backslash\{4\}$	$M\backslash\{4\}$	$\mathbb{C}\backslash\{4\}$	$\mathbb{C}\backslash\{4\}$	$\mathbb{C}\backslash\{4\}$	$\mathbb{C}\backslash\{4\}$
$i=5$	$\varnothing$	$\mathbb{C}\backslash\{5\}$	$\mathbb{C}\backslash\{5\}$	$\mathbb{C}\backslash\{5\}$	$\mathbb{C}\backslash\{5\}$	$M\backslash\{5\}$	$\mathbb{C}\backslash\{5\}$	$\mathbb{C}\backslash\{5\}$	$\mathbb{C}\backslash\{5\}$	$\mathbb{C}\backslash\{5\}$
$i=6$	$\varnothing$	M	M	M	M	M	M	M	M	M

- Illustration of the MmTSPFG:

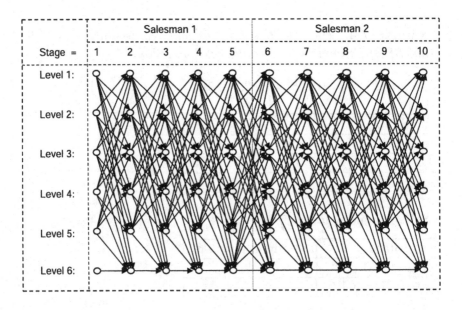

4.4. *Path-based reformulation of BNF-based model*

Definition 7.12 (*"MmTSP path"*). We refer to a path of the MmT-SPFG which spans the set of *stages* of the graph (i.e., a path of length $(n-1)$ of the graph) and which is incident upon each level of the graph pertaining to a customer site in $\mathbb{C}$ at exactly one node of the graph as a *"MmTSP path."* In other words, a set of arcs,

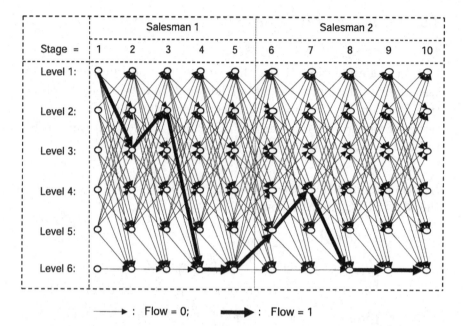

Figure 7.10. Illustration of an *MmTSP path* of the MmTSPFG.

$((i_1, 1, i_2), (i_2, 2, i_3), \dots, (i_{n-1}, n-1, i_n)) \in A^{n-1}$, is a *MmTSP path* iff:

(1) $\forall t \in \mathbb{C}, \exists\, p \in S : i_p = t$; and
(2) $\forall (p, q) \in (S, S \backslash \{p\}) : (i_p, i_q) \in \mathbb{C}^2,\, i_p \neq i_q$.

An illustration of a *MmTSP path* is given in Figure 7.10 for the MmTSP instance of Example 7.3. The *MmTSP path* which is shown on the figure corresponds to the *MmTSP schedule:* $\{(1,1),(1,3), (1,2),(2,5),(2,4)\}$.

Theorem 7.5. *The following statements are true:*

(1) *There exists a one-to-one mapping between the MmTSP paths of the MmTSPFG and the extreme points of the BNF-based Polytope (i.e., the points of P_1);*
(2) *There exists a one-to-one mapping between the MmTSP paths of the MmTSPFG and the MmTSP schedules.*

4.5. *LP model of the MmTSP*

4.5.1. *Objective function "costs"*

Our LP reformulation of the BNF model for the MmTSP consists
essentially of the model stated in Chapter 2. In developing the objective function for the model, we first reformulate the travel cost in
terms of the arcs of the MmTSPFG. Then, we develop the "costs"
to be attached to our modeling variables.

Reformulation of the travel costs. The costs we attach to the arcs of
the MmTSPFG are as follows:

$$\forall r \in R, \forall [i, r, j] \in A,$$

$$\overline{\delta}_{[i,r,j]} := \begin{cases} f_{\mathfrak{p}_r} + e_{\mathfrak{p}_r, \mathfrak{b}_{\mathfrak{p}_r}, i} + e_{\mathfrak{p}_r, i, j} & \text{if } (r = \underline{\mathfrak{r}}_p; \ i \neq \overline{\mathfrak{c}}); \\ 0 & \text{if } ((r = \underline{\mathfrak{r}}_p; \ i = \overline{\mathfrak{c}}) \text{ or } (\overline{\mathfrak{r}}_p = r = n - 1; \ i = j = \overline{\mathfrak{c}})); \\ e_{\mathfrak{p}_r, i, j} & \text{if } ((\underline{\mathfrak{r}}_p < r < \overline{\mathfrak{r}}_p) \text{ or } (\overline{\mathfrak{r}}_p = r < n - 1; \ i = \overline{\mathfrak{c}})); \\ e_{\mathfrak{p}_r, i, \mathfrak{b}_{\mathfrak{p}_r}} & \text{if } ((\overline{\mathfrak{r}}_p = r < n - 1; \ i \in \mathbb{C}) \text{ or} \\ & \quad (\overline{\mathfrak{r}}_p = r = n - 1; \ i \neq \overline{\mathfrak{c}}; \ j = \overline{\mathfrak{c}})); \\ e_{\mathfrak{p}_r, i, j} + e_{\mathfrak{p}_r, j, \mathfrak{b}_{\mathfrak{p}_r}} & \text{if } (\overline{\mathfrak{r}}_p = r = n - 1; \ i \neq \overline{\mathfrak{c}}; \ j \neq \overline{\mathfrak{c}}). \end{cases}$$

"Costs" for the modeling variables. The "cost" we attach to each of
our y-variables is zero. The costs we attach to the z-variables are as
follows:

$$\forall ([i, r, j], [k, s, t], [u, p, v]) \in A^3 : r < s < p,$$

$$\delta_{[i,r,j][k,s,t][u,p,v]}$$
$$:= \begin{cases} \overline{\delta}_{[i,r,j]} + \overline{\delta}_{[k,s,t]} + \overline{\delta}_{[u,p,v]} & \text{if } (r = 1; \ s = 2; \ p = 3); \\ \overline{\delta}_{[u,p,v]} & \text{if } (r = 1; \ s = 2; \ p > 3); \\ 0 & \text{otherwise.} \end{cases}$$

The objective function "costs" are illustrated in Example 7.5.

Example 7.5. *Consider the MmTSP of Example 7.3:*

- *Let the original costs be:*

  - *Salesman "1":*

    - $f_1 = 80$
    - *Inter-site travel costs:*

b_1	1	2	3	4	5
b_1	–	18	16	9	21	15
1	18	–	24	14	14	7
2	4	6	–	21	17	13
3	20	18	3	–	14	28
4	14	27	13	5	–	8
5	29	6	8	16	22	–

  - *Salesman "2":*

    - $f_2 = 90$
    - *Inter-site travel costs:*

b_2	1	2	3	4	5
b_2	–	27	8	5	28	13
1	22	–	21	24	16	11
2	3	11	–	15	14	10
3	18	3	12	–	7	28
4	19	1	17	20	–	6
5	16	24	17	9	20	–

- *The costs to apply to the arcs of the MmTSPFG are illustrated for $i = 4$, $j \in \{3, 6\}$, and $r \in \{1, 2, 5, 9\}$, as follows:*

	$r = 1$	$r = 2$	$r = 5$	$r = 9$
$j = 3$	$80 + 21 + 5 = 106$	5	14	$20 + 18 = 38$
$j = 6$	$80 + 21 + 14 = 115$	14	14	19

4.5.2. *Statement of the LP model of the MmTSP*

Problem 7.7 (Problem MmTSPLP).

Minimize:

$$\sum_{([u,p,v][i,r,j][k,s,t])\in A^3:p<r<s} \delta_{[u,p,v][i,r,j][k,s,t]} z_{[u,p,v][i,r,j][k,s,t]}$$

Subject to:

Constraints (2.1)–(2.9);

$$y_{[i,r,j][k,s,i]} = 0; \quad \forall ([i,r,j],[k,s,t]) \in A^2 :$$

$$((i \neq \mathfrak{c}+1) \ and \ (k=i)) \ or$$

$$((i \neq \mathfrak{c}+1) \ and \ (t=i)) \ or$$

$$((s \neq r+1) \ and \ (j \neq \mathfrak{c}+1) \ and \ (t=j)) \ or$$

$$((j \neq \mathfrak{c}+1) \ and \ (k=j)); \qquad (7.39)$$

$$y_{[i,r,j][k,s,t]} \geq 0, \quad [i,r,j],[k,s,t] \in A; \qquad (7.40)$$

$$z_{[u,p,v][i,r,j][k,s,t]} \geq 0, \quad [u,p,v],[i,r,j],[k,s,t] \in A. \qquad (7.41)$$

As for the SPPLP, constraints (7.39) are the "analogs" of constraints (2.10), and enforce the "no re-visit" condition for the *MmTSP paths* of the MmTSPFG. Note that these constraints apply only to the "actual" customer sites of the MmTSP. As in previous sections of this chapter, constraints (7.40)–(7.41) are the usual non-negativity constraints, and the upper bound constraints on the variables are omitted due to the fact that they are redundant.

Chapter 8

Conclusions

We have presented a generalized framework for formulating hard combinatorial optimization problems as polynomial-sized linear programs. This constitutes a new affirmative resolution of the important "P versus NP" question. Our framework is non-*ad hoc* in that it allows for formulating many of the well-known NP-Complete COP's directly (without the need to reduce them to other COP's) as linear programs.

We have also shown that if one can find an alternate model of a problem which has no need for another model's variables, then these two models are in *independent spaces* and are not necessarily linked as to complexity. Specifically, we have shown that the addition of *redundant* variables and constraints to a model in order to make it project to another model which is in *independent space* can only result in a degenerate, *ill-defined*, ambiguous *extended formulations* relationship between the two models, thus limiting the reach or *barrier* presented by the application of *extended formulations*. In other words, the existence of a one-to-one correspondence between the solutions of one model and the solutions of another model does not necessarily imply that there exists a *well-defined*/non-degenerate/meaningful *extension* relationship between the two models in general. Hence, we believe our developments on *extended formulations* in this book (Chapter 6) brings an important and clearer perspective on the scope of the *extended formulations*-based "*barriers*" to the LP modeling of NP-Complete problems.

Our modeling results are not a suggestion that all problems in the NP complexity class have become equally "easy" to solve, nor that the need for Computational Complexity Theory (CCT) has ceased. However, from a theoretical perspective, we believe that these results make it necessary to reframe the CCT question away from:

"Does there exist a polynomial algorithm for Problem X?"

to (perhaps):

"What is the smallest-dimensional space in which Problem X has a polynomial algorithm?"

The polynomial bound on the size of our proposed generic LP is higher for some COPs and lower for others. Hence, from a practical perspective, algorithmic developments and empirical testing are needed in order to ascertain the computational performance of the models for specific COPs.

Bibliography

Ahuja, R.K., T.L. Magnanti and J.B. Orlin (1993). *Network Flows: Theory, Algorithms, and Applications*. Prentice-Hall, Upper Saddle River, NJ.

Ali, A.I. and H.-S. Han (1998). Reformulation of the set partitioning problem as a pure network with special order set constraints. *Annals of Operations Research* 81:0, pp. 233–249.

Ali, A.I. and J.L. Kennington (1986). The asymmetric *m*-traveling salesmen problem: A duality based branch-and-bound algorithm. *Discrete Applied Mathematics* 13, pp. 259–276.

Ali, A.I. and H. Thiagarajan (1989). A network relaxation based enumeration algorithm for set partitioning. *European Journal of Operational Research* 38:1, pp. 76–85.

Alvarenga, G.B., G.R. Mateus and G. de Tomi (2007). A genetic and set partitioning two-phase approach for the vehicle routing problem with time windows. *Computers & Operations Research* 34:6, pp. 1561–1584.

Balas, E. and M. Padberg (1976). Set partitioning — a survey. *SIAM Review* 18:4, pp. 710–760.

Baldacci, R., N. Christofides and A. Mingozzi (2008). An exact algorithm for the vehicle routing problem based on the set partitioning formulation with additional cuts. *Mathematical Programming* 115:2, pp. 351–385.

Barahona, F. and R. Anbil (2002). On some difficult linear programs coming from set partitioning. *Discrete Applied Mathematics* 118:1/2, pp. 3–11.

Bazaraa, M.S., J.J. Jarvis and H.D. Sherali (2010). *Linear Programming and Network Flows*. Wiley, New York, NY.

Beachy, J.A. and W.D. Blair (2006). *Abstract Algebra*. Waveland Press, Inc., Long Grove, IL.

Bektas, T. (2006). The multiple traveling salesman problem: An overview of formulations and solution procedures. *Omega* 34, pp. 209–219.

Beliën, J. and E. Demeulemeester (2007). On the trade-off between staff-decomposed and activity-decomposed column generation for a staff scheduling problem. *Annals of Operations Research* 155:1, pp. 143–166.

Berger, R.T., C.R. Coullard and M.S. Daskin (2007). Location-routing problems with distance constraints. *Transportation Science* 41:1, pp. 29–43.

Borowski, E.J. and J.M. Borwein (1991). *The* HarperCollins *Dictionary of Mathematics*. HarperCollins Publishers, New York, NY.

Boschetti, M.A., A. Mingozzi and S. Ricciardelli (2008). A dual ascent procedure for the set partitioning problem. *Discrete Optimization* 5:4, pp. 735–747.

Burkard, R.E., M. Dell'Amico and S. Martello (2009). *Assignment Problems*. SIAM, Phildelphia, PA.

Burke, E.K., B. McCollum, A. Meisels, S. Petrovic and R. Qu (2007). A graph-based hyper-heuristic for educational timetabling problems. *European Journal of Operational Research* 176:1, pp. 177–192.

Calamoneri, T. (2006). The $L(h, \mathrm{k})$-labelling problem: A survey and annotated bibliography. *The Computer Journal* 49:5, pp. 585–608.

Cavalcante, V.F., C.C. de Souza and A. Lucena (2008). Relax-and-cut algorithm for the set partitioning problem. *Computers & Operations Research* 35:6, pp. 1963–1981.

Chan, T.J. and C.A. Yano (1992). A multiplier adjustment approach for the set partitioning problem. *Operations Research* 40:1, pp. S40–S47.

Chiang, W.-C. and A.R. Russell (2004). Integrating purchasing and routing in a propane gas supply chain. *European Journal of Operational Research* 154:3, pp. 710–729.

Christofides, N., A. Mingozzi and P. Toth (1981). Algorithms for the vehicle routing problem, based on spanning tree and shortest path relaxations. *Mathematical Programming* 20, pp. 255–282.

Chu, C.-Y., S.B. Tor and G.A. Britton (2008). Graph theoretic algorithm for automatic operation sequencing for progressive die design. *International Journal of Production Research* 46:11, pp. 2965–2988.

Conforti, M., G. Cornuéjols and G. Zambelli (2010). Extended formulations in combinatorial optimization. *Annals of Operation Research* 208:1, pp. 97–143.

Conforti, M., G. Cornuéjols and G. Zambelli (2013). Extended formulations in combinatorial optimization. *Annals of Operations Research* 204:1, pp. 97–143.

Conforti, M., M.D. Summa and G. Zambelli (2007). Minimally infeasible set-partitioning problems with balanced constraints. *Mathematics of Operations Research* 32:3, pp. 497–507.

Coll, P., Marenco, J., Diaz, I.M. and P. Zabala (2002). Facets of the graph coloring polytope. *Annals of Operations Research* 116:1, pp. 79–90.

Coxeter, H.S.M. (1989). *Introduction to Geometry*. Wiley Classics Library Edition. Wiley, Hoboken, NJ.

Cullen, C.G. (1972). *Matrices and Linear Transformations*. Dover Publications, Inc., New York.

Dantzig, G.B., D.R. Fulkerson and S.M. Johnson (1954). Solution of a large-scale traveling salesman problem. *Operations Research* 2:4, pp. 393–410.

de Werra, D. (1985). An Introduction to timetabling. *European Journal of Operational Research* 19:2, pp. 151–162.

Demange, M., T. Ekim and D. de Werra (2009). A tutorial on the use of graph coloring for some problems in robotics. *European Journal of Operational Research* 192:1, pp. 41–55.

Desaulniers, G., A. Langevin, D. Riopel and B. Villeneuve (2003). Dispatching and conflict-free routing of automated guided vehicles: An exact approach. *International Journal of Flexible Manufacturing Systems* 15:4, pp. 309–331.

Desrosiers, J., N. Mladenovic and D. Villeneuve (2005). Design of balanced MBA student teams. *Journal of the Operational Research Society* 56:1, pp. 60–66.

Diaby, M. (2006a). Equality of the complexity classes P and NP: A linear programming formulation of the quadratic assignment problem. *Unpublished*. Available at http://arxiv.org/abs/cs/0609004v4.pdf.

Diaby, M. (2006b). On the Equality of Complexity Classes P and NP: Linear Programming Formulation of the Quadratic Assignment Problem. *Proceedings of the IMECS 2006*, Hong Kong, China, pp. 774–779.

Diaby, M. (2007). The traveling salesman problem: A linear programming formulation. *WSEAS Transactions on Mathematics* 6:6, pp. 745–754.

Diaby, M. (2008). A $O(n8) \times O(n7)$ linear programming model of the traveling salesman problem. *Unpublished*. Available at http://arxiv.org/abs/0803.4354.

Diaby, M. (2010a). Linear programming formulation of the multi-depot multiple traveling salesman problem with differentiated travel costs. In D. Davendra (ed.), *Traveling Salesman Problem, Theory and Applications*. InTech, New York, NY, pp. 257–282.

Diaby, M. (2010b). Linear programming formulation of the set partitioning problem. *International Journal of Operational Research* 8:4, pp. 399–427.

Diaby, M. (2010c). Linear programming formulation of the vertex coloring problem. *International Journal of Mathematics in Operational Research* 2:3, pp. 259–289.

Diaby, M. and M.H. Karwan (2015). Limits to the scope of applicability of extended formulations theory for LP models of combinatorial optimisation problems. *International Journal of Mathematics in Operational Research* (Forthcoming).

Dowsland, K.A. and J.M. Thompson (2005). Ant colony optimization for the examination scheduling problem. *Journal of the Operational Research Society* 56:4, pp. 426–438.

Edmonds, J. (1970). Submodular functions, matroids and certain polyhedra". In: R.K. Guy *et al.* (eds.), *Combinatorial Structures and Their Applications.* Gordon and Breach, New York, pp. 69–87.

El-Darzi, E. and G. Mitra (1992). Solution of set-covering and set-partitioning problems using assignment relaxations. *Journal of the Operational Research Society* 43:5, pp. 483–493.

El-Darzi, E. and G. Mitra (1995). Graph theoretical relaxations of set covering and set partitioning problems. *European Journal of Operational Research* 87:1, pp. 109–121.

Eveborn, P., P. Flisberg and M. Ronnqvist (2006). L aps C are — an operational system for staff planning of home care. *European Journal of Operational Research* 171:3, pp. 962–976.

Eveborn, P. and M. Ronnqvist (2004). Scheduler — a system for staff planning. *Annals of Operations Research* 128:1–4, pp. 21–45.

Fei, H., C. Chu, N. Meskens and A. Artiba (2008). Solving surgical cases assignment problem by a branch-and-price approach. *International Journal of Production Economics* 112:1, pp. 96–108.

Fiorini, S., S. Massar, S. Pokutta, H.R. Tiwary and R. de Wolf (2011). Linear vs. semidefinite extended formulations: Exponential lower bounds for polytopes in combinatorial Optimization. *Unpublished.* Available at http://arxiv.org/pdf/1111.0837.pdf. (Last revised March 2015).

Fiorini, S., S. Massar, S. Pokutta, H.R. Tiwary and R. de Wolf (2012). Linear vs. semidefinite extended formulations: Exponential separation and strong bounds. *Proceedings of the 44th ACM Symposium on the Theory of Computing (STOC'12)*, New York, NY, pp. 95–106.

Fisher, M.L. and P. Kedia (1990). Optimal solution of set covering/ partitioning problems using dual heuristics. *Management Science* 36:6, pp. 674–688.

Freling, R., D. Huismanand and A.P.M. Wagelmans (2003). Models and algorithms for integration of vehicle and crew scheduling. *Journal of Scheduling* 6:1, pp. 63–85.

Fritsch, R. and G. Fritsch (1998). *The Four-Color Theorem.* Wiley, New York.

Galinier, P. and J.K. Hao (1999). Hybrid evolutionary algorithms for graph coloring. *Journal of Combinatorial Optimization* 3:4, pp. 379–397.

Galinier, P. and A. Hetz (2006). A survey of local search methods for graph coloring. *Computers and Operations Research* 33:9, pp. 2547–2562.

Gamache, M., A. Hertz and J.O. Ouellet (2007). A graph coloring model for a feasibility problem in monthly crew scheduling with preferential bidding. *Computers & Operations Research* 34:8, pp. 2384–2395.

Gamelin, T.W. and R.E. Greene (1999). *Introduction to Topology*. Dover Publications, Inc., Mineola, NY.

Garey, M.R. and D.S. Johnson (1979). *Computers and Intractability: A Guide to the Theory of NP-Completeness*. Freeman and Company, New York, NY.

Garey, M.R., D.S. Johnson and R.E. Tarjan (1976). The planar Hamiltonian circuit problem is NP-Complete. *SIAM Journal on Computing* 5:4, pp. 704–714.

Gavish, B. and K. Srikanth (1986). An optimal solution method for large-scale multiple traveling salesman problems. *Operations Research* 34, pp. 698–717.

Gebremedhin, A.H., F. Manne and A. Pothen (2005). What color is your Jacobian? Graph coloring for computing derivatives. *SIAM Review* 47:4, pp. 629–705.

Ghafurian, S. and N. Javadian (2011). An ant colony algorithm for solving fixed destination multi-depot multiple traveling salesman problem. *Applied Soft Computing* 11:March, pp. 1256–1262.

Gomez, D., J. Montero, J. Yanez and C. Poidomani (2007). A graph coloring approach for image segmentation. *Omega* 35:2, pp. 173–183.

Govindaraju, N.K., D. Knott, N. Jain, I. Kabul, R. Tamstorf, R. Gayle, M.C. Lin and D. Manocha (2005). Interactive collision detection between deformable models using chromatic decomposition. *ACM Transactions on Graphics* 24:3, pp. 991–999.

Grigoriev, A., J.V.D. Klundert and F.C.R. Spieksma (2006). Modeling and solving the periodic maintenance problem. *European Journal of Operational Research* 172:3, pp. 783–797.

Gromicho, J., J. Paixã and I. Branco (1992). Exact solution of multiple traveling salesman problems. In M. Akgül *et al.* (eds.), *Combinatorial Optimization*. Springer, Berlin.

Harche, F. and G.L. Thompson (1994). The column subtraction algorithm: An exact method for solving weighted set covering, packing and partitioning problems. *Computers & Operations Research* 21:6, pp. 689–705.

Hoffman, K.L. and M. Padberg (1993). Solving airline crew scheduling problems by branch-and-cut. *Management Science* 39:6, pp. 657–682.

Hofman, R. (2006). Report on article: P=NP: Linear programming formulation of the traveling salesman problem. *Unpublished*. Available at http://arxiv.org/abs/cs/0610125.

Hofman, R. (2007). Why linear programming cannot solve large instances of NP-complete problems in polynomial time. *Unpublished*. Available at http://arxiv.org/abs/cs/0611008.

Hofman, R. (2008). Report on article: The traveling salesman problem: A linear programming formulation. *Unpublished.* Available at http:// arxiv.org/abs/0805.4718.

Hong, S.-P., K.M. Kim, K. Lee and B.H. Park (2009). A pragmatic algorithm for the train-set routing: The case of Korea high-speed railway. *Omega* 37:3, pp. 637–645.

Ileri, Y., M. Bazaraa, T. Gifford, G.L. Nemhauser, J. Sokol and E. Wilkum (2006). An optimization approach for planning daily drayage operations. *Central European Journal of Operations Research* 14:2, pp. 141–156.

Jepsen, M., B. Petersen, S. Spoorendonk and D. Pisinger (2008). Subset-row inequalities applied to the vehicle-routing problem with time windows. *Operations Research* 56:2, pp. 497–513.

Joseph, A. (2002). A concurrent processing framework for the set partitioning problem. *Computers & Operations Research* 29:10, pp. 1375–1391.

Jünger, M., T. Liebling, D. Naddef, G. Nemhauser, W. Pulleyblank, G. Reinelt, G. Rinaldi and L. Wolsey (eds.) (2010). *50 Years of Integer Programming 1958–2008.* Springer, New York, NY.

Jyotheswar, J. and S. Mahapatra (2007). Efficient FPGA implementation of DWT and modified SPIHT for lossless image compression. *Journal of Systems Architecture* 53:7, pp. 369–378.

Kaibel, V. (2011). Extended formulations in combinatorial optimization. *Optima* 85:2, pp. 2–7.

Kaibel (2013). *Private email communications on:* "LP-formulations that induce extensions" by V. Kaibel, M. Walter, and S. Weltge.

Kania, D. and J. Kulisz (2007). Logic synthesis for PAL-based CPLD-s based on two-stage decomposition. *The Journal of Systems and Software* 80:7, pp. 1129–1141.

Karp, R.M. (1972). Reducibility among combinatorial problems. In R.E. Miller and J.W. Thatcher (eds.), *Complexity of Computer Computations.* Plenum Press, New York, NY, pp. 85–103.

Kinney, G.W., J.W. Barnes and B.W. Colletti (2007). A reactive Tabu search algorithm with variable clustering for the unicost set covering problem. *International Journal of Operational Research* 2:2, pp. 156–172.

Klabjan, D. (2004). A practical algorithm for computing a subadditive dual function for set partitioning. *Computational Optimization and Applications* 29:3, pp. 347–368.

Kliewer, N., T. Mcllouli and L. Suhl (2006). A time-space network based exact optimization model for multi-depot bus scheduling. *European Journal of Operational Research* 175:3, pp. 1616–1627.

Kochenberger, G.A., F. Glover, B. Alidaee and C. Rego (2005). An unconstrained quadratic binary programming approach to the vertex coloring problem. *Annals of Operations Research* 139:1, pp. 229–241.

Laguna, M. and R. Marti (2001). A GRASP for coloring sparse graphs. *Computational Optimization and Applications* 19:2, 165–178.

Laporte, G. and Y. Nobert (1980). A cutting planes algorithm for the *m*-salesmen problem. *Journal of the Operational Research Society* 31, pp. 1017–1023.

Lawler, E.L., J.K. Lenstra, A.H.G. Rinnooy Kan and D.B. Shmoys (eds.) (1985). *The Traveling Salesman Problem: A Guided Tour of Combinatorial Optimization.* Wiley, New York, NY.

Lee, Y.H., J.I. Kim, K.H. Kang and K.H. Kim (2008). A heuristic for vehicle fleet mix problem using tabu search and set partitioning. *Journal of the Operational Research Society* 59:6, pp. 833–841.

Lewis M., G. Kochenberger and B. Alidaee (2008). A new modeling and solution approach for the set-partitioning problem. *Computers & Operations Research* 35:3, pp. 807–813.

Linderoth, J.T., E.K. Lee and M.W.P. Savelsbergh (2001). A parallel, linear programming-based heuristic for large-scale set partitioning problems. *INFORMS Journal on Computing* 13:3, pp. 191–209.

Lucena, A. (2005). Non delayed relax-and-cut algorithms. *Annals of Operations Research* 140:1, pp. 375–410.

Magaril-Il'yaev, G.G. and V.M. Tikhomirov (2000). *Convex Analysis: Theory and Applications.* American Mathematical Society, Providence, RI.

Mahdavi, I., J. Rezaeian, K. Shanker and Z.R. Amiri (2006). A set partitioning based heuristic procedure for incremental cell formation with routing flexibility. *International Journal of Production Research* 44:24, pp. 5343–5361.

Malaguti, E., M. Monaci and P. Toth (2008). A metaheuristic approach for the vertex coloring problem. *INFORMS Journal on Computing* 20:2, pp. 302–316.

Marsten, R.E. (1974). An algorithm for large set partitioning problems. *Management Science* 20:5, pp. 774–787.

Marsten, R.E. and F. Shepardson (1981). Exact solution of crew scheduling problems using the set partitioning model: Recent successful applications. *Networks* 11:2, pp. 165–177.

Martin, R.K. (1991). Using separation algorithms to generate mixed integer model reformulations. *Operations Research Letters* 10:3, pp. 119–128.

Medard, C.P. and N. Sawhney (2007). Airline crew scheduling from planning to operations. *European Journal of Operational Research* 183:3, pp. 1013–1027.

Mesquita, M. and A. Paias (2008). Set partitioning/covering-based approaches for the integrated vehicle and crew scheduling problem. *Computers & Operations Research* 35:5, pp. 1562–1575.

Miller, C.E., A.W. Tucker and R.A. Zemlin (1960). Integer programming formulations and traveling salesman problems. *Journal of the Association for Computing Machinery* 7:4, pp. 326–329.

Minkowski, H. (1910). *Geometrie der Zahlen.* Teubner, Leipzig.

Nemhauser, G.L. and L.A. Wolsey (1988). *Integer and Combinatorial Optimization.* Wiley, New York, NY.

Noronha, T.F. and C.C. Ribeiro (2006). Routing and wavelength assignment by partition colouring. *European Journal of Operational Research* 171:3, pp. 797–810.

Oliveira, C.A.S., P.A. Pardalos and T.M. Querido (2005). A combinatorial algorithm for message scheduling on controller area networks. *International Journal of Operational Research* 1:1/2, pp. 160–171.

Öncan, T., K. Altinel and G, Laporte (2009). A comparative analysis of several asymmetric traveling salesman problem formulations. *Computers & Operations Research* 36:3, pp. 637–654.

Osei-Bryson, K.-M. and A. Joseph (2006). Applications of sequential set partitioning: A set of technical information systems problems. *Omega* 34:5, pp. 492–500.

Panik, M.J. (1993). *Fundamentals of Convex Analysis.* Kluwer Academic Publishers, Boston, MA.

Papadimitriou, C.H. and K. Steiglitz (1982). *Combinatorial Optimization: Algorithms and Complexity.* Prentice-Hall, Englewood Cliffs.

Pardalos, P.M., T. Mauridon and J. Xue (1999). The graph coloring problem: A bibliographic survey. In D.Z. Du and P.M. Pardalos (eds.), *Handbook of Combinatorial Optimization*, Vol. 2. Kluwer Academic Publishers, Dordrecht, Holland, pp. 331–395.

Peng, C., H. Zheng and B.Y. Zhao (2006). Utilization and fairness in spectrum assignment for opportunistic spectrum access. *Mobile Networks and Applications* 11:4, pp. 555–576.

Regan, K.W. (2015). *Private Email Communications and Discussions.*

Regan, K.W. and R.J. Lipton (2013). *Private Email Communications.*

Rockafellar, R.T. (1997). *Convex Analysis.* Princeton University Press, Princeton.

Sadler, A. and C. Gervet (2008). An exact algorithm for a cross-docking supply chain network design problem. *Journal of Heuristics* 14:1, pp. 23–67.

Sarin, S.C., H.D. Sherali and A. Bhootra (2005). New tighter polynomial length formulations for the asymmetric travelling salesman problem with and without precedence constraints. *Operations Research Letters* 33:1, pp. 62–70.

Saygin, C., K. Cha, M. Zawodniok, A. Ramachandran and J. Sarangapani (2006). Interference mitigation and read rate improvement in RFID-based network-centric environments. *Sensor Review* 26:4, pp. 318–325.

Schrijver, A. (1986). *Theory of Linear and Integer Programming*. Wiley, New York, NY.

Sindhuchao, S., H.E. Romeijn, E. Akçali and R. Boondiskulchok (2005). An integrated inventory-routing system for multi-item joint replenishment with limited vehicle capacity. *Journal of Global Optimization* 32:1, pp. 93–118.

Swart, E.R. (1986; 1987 revision). "P=NP". *Technical Report*. University of Guelph, Canada.

Tang, L., G. Wang and J. Liu (2007). A branch-and-price algorithm to solve the molten iron allocation problem in iron and steel industry. *Computers & Operations Research* 34:10, pp. 3001–3015.

Teo, C.-P. and J. Shu (2004). Warehouse–retailer network design problem. *Operations Research* 52:3, pp. 396–408.

Thomadsen, T. and J. Larsen (2007). A hub location problem with fully interconnected backbone and access networks. *Computers & Operations Research* 34:8, pp. 2520–2531.

Tombus, O. and T. Bilgic (2004). A column generation approach to the coalition formation problem in multi-agent systems. *Computers & Operations Research* 31:10, pp. 1635–1653.

Vanderbeck, F. and L.A. Wolsey (2010). Reformulation and decomposition of integer programs. In Jünger *et al.* (eds.), *50 Years of Integer Programming 1958-2008*, pp. 431–502. Springer, New York, NY.

Westphal, S. and S.O. Krumke (2008). Pruning in column generation for service vehicle dispatching. *Annals of Operations Research* 159:1, pp. 355–371.

Weyl, H. (1935). Elementare theorie der konvexen polyheder. *Commentarii Math. Helvetici* 7, pp. 290–306.

Woo, T.K., S.Y.W. Su and R.N. Wolfe (2002). Resource allocation in a dynamically partitionable bus network using graph coloring algorithm. *IEEE Transactions on Communications* 39:12, pp. 1794–1801.

Yannakakis, M. (1991). Expressing combinatorial optimization problems by linear programming. *Journal of Computer and System Sciences* 43:3, pp. 441–466.

Yannakakis, M. (2013). *Private Email Communications*.

Zufferey, N., P. Amstutz and P. Giaccari (2008). Graph colouring approaches for a satellite range scheduling problem. *Journal of Scheduling* 11:4, pp. 263–277.

Appendix A

On the (Two) Counter-Example Claims

1. Preliminaries: Shortfall of the Model Without z-Variables

The model presented in this section was the first focus of our research efforts. Because our empirical testing revealed its non-integrality, we did not pursue its development beyond the attempts at the very early stages of the research. The model and its relaxation which will be described in this appendix are important because they are the bases of all of the counter-example claims that we know of (through public media (Hofman (2006; 2008), and also through anonymous reviews) in relation to our modeling approach, as briefly discussed in the introduction chapter (Chapter 1) of this book. We will discuss, in particular, why the proof logic used in Chapter 3 of this book is not applicable to the model to be discussed below, in this section.

For the purpose of the discussions in this section, we introduce a new class of variables defined in the following notation:

Notation A.1. $\forall r \in R$, $\forall (i, j) \in (N_r, F_r(i))$, *we denote the total amount of flow in the TSPFG that traverses arc* $[i, r, j]$, *as* $x_{[i,r,j]}$.

1.1. "Full" constraint set

The model is stated in the form of those that are the subject of Hofman (2006; 2008) for convenience. In this statement, any variable that is not defined according to Notation A.1, should be assumed to be equal to zero. The model is as follows:

$$\sum_{i \in M} \sum_{j \in M} \sum_{t \in M} y_{[i,1,j][j,2,t]} = 1, \tag{A.1}$$

$$\sum_{v \in M} y_{[i,r,j][v,s-1,t]} - \sum_{v \in M} y_{[i,r,j][t,s,v]} = 0;$$
$$i,j,t \in M; \quad r,s \in R : r < s-1, \tag{A.2}$$

$$\sum_{v \in M} y_{[v,s-1,t][i,r,j]} - \sum_{v \in M} y_{[t,s,v][i,r,j]} = 0;$$
$$i,j,t \in M; \quad r,s \in R : 1 < s < r, \tag{A.3}$$

$$x_{[i,r,j]} - \sum_{u \in M} \sum_{v \in M} y_{[u,p,v][i,r,j]} = 0;$$
$$i,j \in M; \quad r,p \in R : p < r, \tag{A.4}$$

$$x_{[i,r,j]} - \sum_{u \in M} \sum_{v \in M} y_{[i,r,j][u,p,v]} = 0;$$
$$i,j \in M; \quad r,p \in R : p > r, \tag{A.5}$$

$$x_{[i,r,j]} - \sum_{p \in R: p<r} \sum_{v \in M} y_{[t,p,v][i,r,j]} - \sum_{p \in R: p>r} \sum_{v \in M} y_{[i,r,j][v,p,t]} = 0;$$
$$i,j,t \in M : i \neq j \neq t; \quad r \in R, \tag{A.6}$$

$$\sum_{(k,t) \in M^2} y_{[i,r,j][k,r+1,t]} = 0; \quad i,j \in M; \quad r \in R \setminus \{n-2\}, \tag{A.7}$$

$$\sum_{s \in R: \, s>r} \sum_{k \in M} \sum_{t \in M} y_{[i,r,i][k,s,t]} + \sum_{s \in R: \, s<r} \sum_{k \in M} \sum_{t \in M} y_{[k,s,t][j,r,j]}$$
$$+ \sum_{s \in R: s \geq r+1} \sum_{k \in M} y_{[i,r,j][k,s,i]} + \sum_{s \in R: s>r} \sum_{k \in M} y_{[i,r,j][i,s,k]}$$
$$+ \sum_{s \in R: s>r} \sum_{k \in M} y_{[i,r,j][k,s,j]} + \sum_{s \in R: s>r+1} \sum_{k \in M} y_{[i,r,j][j,s,k]} = 0;$$
$$i,j \in M; \quad r \in R, \tag{A.8}$$

$$x_{[i,r,j]} \geq 0, \quad i,j \in M; \quad r \in R; \quad y_{[i,r,j][k,s,t]} \geq 0,$$
$$i,j,k,\, t \in M; \quad r,s \in R : r < s. \tag{A.9}$$

Notation A.2. *We denote the polytope induced by constraints* (A.1)–(A.9) *as* Y_L.

In order to facilitate the discussion, we re-express some of the "*communication*" notions in Chapter 3 of this book in terms of the points of Y_L in the following definition.

Definition A.1. Let $(x, y) \in Y_L$:

(1) Three arcs of the TSPFG are said to 3-*communicate* in (x, y) iff each pair of arcs in the triplet 2-*communicate*.

   In other words, we will say that arcs (a), (b), and (c) of the TSPFG 3-*communicate* in (x, y) iff: (a) and (b) 2-*communicate* in (x, y); (a) and (c) 2-*communicate* in (x, y); and (b) and (c) 2-*communicate* in (x, y).

(2) We re-express the notion of a "*communication path*" as follows. For $(r, s) \in R^2 : s > r$, we refer to the set of arcs, $\{[i_r, r, j_r], \ldots, [i_s, s, j_s]\}$, of the TSPFG as a "*y-communication path* from $[i_r, r, j_r]$ to $[i_s, s, j_s]$" iff $\forall (p, q) \in R^2 : r \leq p < q \leq s$, $y_{[i_p, p, j_p][i_q, q, j_q]} > 0$.

In order to apply the proof logic of Chapter 3 of this book to Y_L one must show that for a given solution instance of Y_L there exists at least one *y-communication path* of the solution linking any two arcs of the TSPFG that 2-*communicate* in the solution. The difficulty in this respect however, is with the inductive step ("Case 4") of Theorem 3.4 of Chapter 3 of this book. This is due to the following fact:

Remark A.1.

It is possible for two given arcs at stages r and s $(s > r + 3)$ of the TSPFG, respectively, to 2-*communicate* in a given solution instance of Y_L, and for there not to exist an arc at stage $r + 1$ of the TSPFG that 3-*communicates* in the solution with the two given arcs.

   Similarly, it is possible for two given arcs at stages r and s $(s > r + 3)$ of the TSPFG, respectively, to 2-*communicate* in a given solution instance of Y_L, and for there not to exist an arc at stage $s - 1$ of the TSPFG that 3-*communicates* in the solution with the two given arcs.

In other words, let $(x, y) \in Y_L$. Then, the following are true:

(1) For $([i,r,j], [k,s,t]) \in A^2 : s > r+3$,

$$y_{[i,r,j][k,s,t]} > 0$$
$$\Rightarrow \left(\exists \langle u \in N_{r+2}\rangle : \langle y_{[i,r,j][j,r+1,u]} > 0 \text{ and } y_{[j,r+1,u][k,s,t]} > 0\rangle\right);$$

(2) For $([i,r,j], [k,s,t]) \in A^2 : s > r+3$,

$$y_{[i,r,j][k,s,t]} > 0$$
$$\Rightarrow \left(\exists \langle v \in N_{s-1}\rangle : \langle y_{[i,r,j][v,s-1,k]} > 0 \text{ and } y_{[v,s-1,k][k,s,t]} > 0\rangle\right).$$

Statement (1) of Remark A.1 is illustrated in Figure A.1. For the sake of clarity, the solution presented on the figure is only a partial one (since it does not span the set of stages of the TSPFG), although

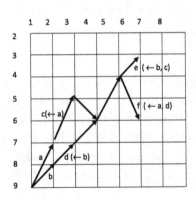

- "$\leftarrow$": *"receives flow from"*
- Each (non-zero) y-variable is equal to 0.5;
- $(y_{(9,1,7)(7,2,5)} > 0;\ y_{(9,1,7)(4,5,6)} > 0;\ y_{(7,2,5)(4,5,6)} = 0) \Rightarrow$
  $y_{(9,1,7)(4,5,6)} > 0$ and $(\nexists\langle u \in M\rangle: \langle y_{(9,1,7)(7,2,u)} > 0$ and $y_{(7,2,u)(4,5,6)} > 0\rangle);$
- $(y_{(9,1,8)(8,2,7)} > 0;\ y_{(9,1,8)(4,5,3)} > 0;\ y_{(8,2,7)(4,5,3)} = 0) \Rightarrow$
  $y_{(9,1,8)(8,2,7)} > 0$ and $(\nexists\langle u \in M\rangle: \langle y_{(9,1,8)(8,2,u)} > 0$ and $y_{(8,2,u)(4,5,3)} > 0\rangle).$

Figure A.1. Illustration of Remark A.1.

it is easy to verify that it can be augmented (with relative ease) into a "complete solution" that satisfies each of the constraints of Y_L.

In the example of Figure A.1, arc $[9, 1, 7]$ (labeled as "(a)") *2-communicates* with arc $[4, 5, 6]$ (labeled as "(f)"), but there exists no arc at Stage 2 of the graph which *3-communicates* with these two arcs (i.e., $[9, 1, 7]$ and $[4, 5, 6]$). Similarly, arc $[9, 1, 8]$ of the example (labeled as "(b)") *2-communicates* with arc $[4, 5, 3]$ (labeled as "(e)"), but there exists no arc at Stage 2 of the graph which *3-communicates* with these two arcs ($[9, 1, 8]$ and $[4, 5, 3]$).

A direct consequence of Remark A.1 with respect to the developments in Chapter 3 of this book is that it implies that conditions (i) and (ii) of "Remark 26.4" in Chapter 3 of this book do not necessarily hold for a point of Y_L. A consequence of this in turn, is that the inductive step ("Case 4") of Theorem 3.4 cannot be applied.

In reference to the illustrative examples on Figures 3.20 and 3.21 of Chapter 3 of this book for example, because of Remark A.1, it is not necessary in the case of $(x, y) \in Y_L$ that each of the arcs that would correspond to the "*u-arcs*" in these illustrative examples *2-communicate* with the *destination arc* (Point #1 of the examples). Similarly, it is not necessary that each of the arcs that would correspond to the "*v-arcs*" of the illustrative examples will *2-communicate* (in (x, y)) with the *source arc* (Point #2 of the examples). Hence, neither Point #1 nor Point #2 shown in the illustrative examples is true in general for $(x, y) \in Y_L$. Hence, the proof logic in Chapter 3 of this book cannot be applied to Y_L.

1.1.1. *"Visit-relaxed" model*

In this section, we consider a version of the model (A.1)–(A.9) in which constraints (A.6) are removed (i.e., $\{(x, y) \in \mathbb{R}^{(\xi_x + \xi_y)} : (x, y)$ satisfies (A.1)–(A.5) and (A.7)–(A.9)$\}$, where: $\xi_x := m(m - 1)^2$). We present a simple counter-example that shows the non-integrality of this relaxed model.

Let the support subgraph of the TSPFG for a 9-city TSP be as shown in Figure A.2. Consider the flow pattern over this subgraph depicted in Figures A.3 and A.4 for the arcs at Stages 1 and 4

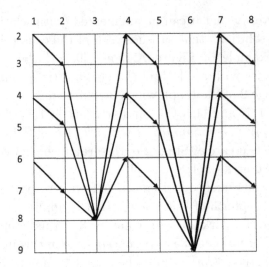

Figure A.2. Support subgraph of a feasible solution to the relaxed model.

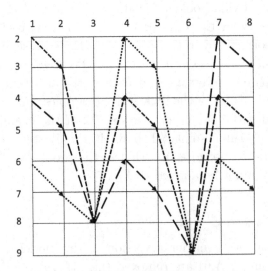

Figure A.3. Propagation of Stage 1 flows.

respectively. (The flow propagations from other stages are trivial to infer and are therefore not depicted separately.)

Then, it can be easily verified that a feasible solution to the relaxed model is obtained by setting each of the y-variables defined by this flow pattern to a value of $1/3 = 0.\overline{33}$. However, clearly, such a

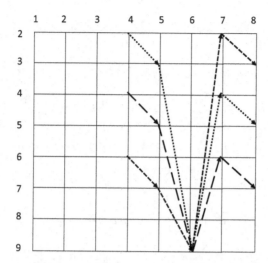

Figure A.4. Propagation of Stage 4 flows.

solution does not correspond to a convex combination of TSP tours
(as would be characterized in terms of *y-communication paths*, consistent with the developments in Chapter 3 of this book).

Hence, the relaxed polytope $\{(x,y) \in \mathbb{R}^{(\xi_x + \xi_y)} : (x,y)$ satisfies
(A.1)–(A.5) and (A.7)–(A.9)$\}$ cannot be integral in general.

2. Counter-Example Claim #1 (Hofman (2006))

The counter-example claim in Hofman (2006) is with respect to the
relaxation of the model in Diaby (2006b) suggested in Diaby (2006a).
We will first state the equivalent of the model Diaby (2006b) in terms
of the notations used in this book (see Notations 1.1 and 2.1). Then,
we will discuss the relaxation suggested in Diaby (2006b).

2.1. The "full" constraint set (Diaby, 2006b)

$$\sum_{i \in N_r} \sum_{j \in F_r(i)} \sum_{t \in F_{r+1}(j)} y_{[i,1,j][j,2,t]} = 1; \qquad (A.10)$$

$$\sum_{v \in B_s(t)} y_{[i,r,j][v,s-1,t]} - \sum_{v \in F_s(t)} y_{[i,r,j][t,s,v]} = 0;$$

$$([i,r,j] \in A; \ (t,s) \in \overline{N}) : r < s - 1; \qquad (A.11)$$

$$\sum_{v \in B_s(t)} y_{[v,s-1,t][i,r,j]} - \sum_{v \in F_s(t)} y_{[t,s,v][i,r,j]} = 0;$$

$$([i,r,j] \in A; \ (t,s) \in \overline{N}) : s < r; \qquad (A.12)$$

$$y_{[i,r,j][k,s,t]} - \sum_{u \in N_r} \sum_{v \in F_p(u)} z_{[i,r,j][k,s,t][u,p,v]} = 0;$$

$$([i,r,j], \ [k,s,t] \in A, \ p \in R) : r < s < p, \qquad (A.13)$$

$$y_{[i,r,j][k,s,t]} - \sum_{u \in N_r} \sum_{v \in F_p(u)} z_{[i,r,j][u,p,v][k,s,t]} = 0;$$

$$([i,r,j], \ [k,s,t] \in A, \ p \in R) : r < p < s, \qquad (A.14)$$

$$y_{[i,r,j][k,s,t]} - \sum_{u \in N_r} \sum_{v \in F_p(u)} z_{[u,p,v][i,r,j][k,s,t]} = 0;$$

$$([i,r,j], \ [k,s,t] \in A, \ p \in R) : p < r < s, \qquad (A.15)$$

$$y_{[i,r,j][k,s,t]} - \sum_{p \in R: p < r} \sum_{v \in F_p(u)} z_{[u,p,v][i,r,j][k,s,t]}$$

$$- \sum_{p \in R: r < p < s} \sum_{v \in B_p(u)} z_{[i,r,j)][v,p,u][k,s,t]}$$

$$- \sum_{p \in R: p > s} \sum_{v \in B_p(u)} z_{[i,r,j][k,s,t][v,p,u]} = 0;$$

$$[i,r,j], \ [k,s,t] \in A : \ r < s, \ u \in M \backslash \{i,j,k,t\}, \quad (A.16)$$

$$y_{[i,r,j][k,s,t]} = 0; \ [i,r,j], \ [k,s,t] \in A:$$

$$(r < s; \ (s = r + 1 \text{ and } k \neq j)), \qquad (A.17)$$

$$y_{[i,r,j][k,s,t]} = 0; \ [i,r,j], \ [k,s,t] \in A:$$

$$(r < s < p; \ (i,j,k,t) \notin \overline{C}(r,s,p)), \qquad (A.18)$$

$$y_{[i,r,j][k,s,t]}, \ z_{[i,r,j][k,s,t][u,p,v]} \geq 0,$$

$$i,j,k,t,u,v \in M; \ r,s,p \in R : r < s < p, \qquad (A.19)$$

where:

$$\forall (r,s) \in R^2 : r < s,$$

$$\overline{C}(r,s) := \left(\begin{array}{l} ([i,r,j],[k,s,t]) \in A^2 : \\ (i \neq j \neq t \quad \text{if } s = r+1; \\ i \neq j \neq k \neq t \quad \text{if } s > r+1). \end{array} \right).$$

2.2. *The relaxed model*

The relaxation suggested in Diaby (2006a) ("Proposition 6" on p. 20) that is the subject of the claim in Hofman (2006) consists of the model above with constraints (A.16) relaxed. Because the objective function in Diaby (2006a; 2006b) involves only the y-variables, the suggested relaxation separates in terms of the y- and z-variables, with the consequence of rendering the z-variables and constraints (A.13)–(A.15) *redundant* in the model. Because of this, the suggested relaxation reduces to the "visit-relaxed" model discussed in Section 1 of this appendix. Hence, the relaxation suggested in Diaby (2006a) cannot be integral in general. However, this non-integrality is not due to problem size, contrary to the claim in Hofman (2006).

3. Counter-Example Claim #2 (Hofman (2008))

The model that is the subject of this section is a relaxation of the model developed in Diaby (2007b) discussed in Diaby (2008). The relaxation consists of dropping all of the z-variables that have their first arcs (in the arc triplet of indices) originating from stages other than the first stage. This relaxed model is as follows:

$$\sum_{i\in M}\sum_{j\in M}\sum_{v\in M}\sum_{t\in M} z_{[i,1,j][j,2,v][v,3,t]} = 1, \qquad (A.20)$$

$$\sum_{v\in M} z_{[i,1,j][k,s,t][v,p,u]} - \sum_{v\in M} z_{[i,1,j][k,s,t][u,p+1,v]} = 0;$$

$$i,j,k,t,u \in M; \quad p,s \in R : 1 < s < p < n-2, \quad (A.21)$$

$$\sum_{v \in M} z_{[i,1,j][v,p,u][k,s,t]} - \sum_{v \in M} z_{[i,1,j][u,p+1,v][k,s,t]} = 0;$$

$$i,j,k,t,u \in M; \quad p,s \in R : 1 < p < s-1, \ s > 3, \quad \text{(A.22)}$$

$$y_{[i,1,j][u,p,v]} - \sum_{k \in M} \sum_{t \in M} z_{[i,1,j][u,p,v][k,s,t]} = 0;$$

$$i,j,u,v \in M; \quad p,s \in R : 1 < p < s, \quad \text{(A.23)}$$

$$y_{[i,1,j][k,s,t]} - \sum_{u \in M} \sum_{v \in M} z_{[i,1,j][u,p,v][k,s,t]} = 0;$$

$$i,j,k,t \in M; \quad p,s \in R : 1 < p < s, \quad \text{(A.24)}$$

$$y_{[i,1,j][k,s,t]} - \sum_{p \in R:1<p<s} \sum_{v \in M} z_{[i,1,j][u,p,v][k,s,t]}$$

$$- \sum_{p \in R:p>s} \sum_{v \in M} z_{[i,1,j][k,s,t][v,p,u]} = 0; \ i,j,k,t,u \in M; \ s \in R\backslash\{1\},$$

$$\text{(A.25)}$$

$$y_{[u,p,v][k,s,t]} - \sum_{i \in M} \sum_{j \in M} z_{[i,1,j][u,p,v][k,s,t]} = 0;$$

$$u,v,k,t \in M; \quad p,s \in R, \ 1 < p < s, \quad \text{(A.26)}$$

$$\sum_{(k,t)\in M^2} y_{[i,r,j][k,r,t]} + \sum_{\substack{(k,t)\in M^2: \\ k\neq j; \ (k,r+1,t)\in A}} y_{[i,r,j][k,r+1,t]} = 0;$$

$$i,j \in M; \quad r \in R, \quad \text{(A.27)}$$

$$\sum_{s \in R: \ s>r} \sum_{k \in M} \sum_{t \in M} y_{[i,r,i][k,s,t]}$$

$$+ \sum_{s \in R: \ s<r} \sum_{k \in M} \sum_{t \in M} y_{[k,s,t][j,r,j]} \sum_{s \in R:s\geq r+1} \sum_{k \in M} y_{[i,r,j][k,s,i]}$$

$$+ \sum_{s \in R:s\geq r+1} \sum_{k \in M} y_{[i,r,j][i,s,k]} + \sum_{s \in R:s\geq r+1} \sum_{k \in M} y_{[i,r,j][k,s,j]}$$

$$+ \sum_{s \in R:s\geq r+2} \sum_{k \in M} y_{[i,r,j][j,s,k]} = 0; \quad i,j \in M; \quad r \in R, \quad \text{(A.28)}$$

$$y_{[i,r,j][k,s,t]} \geq 0; \quad i,j,k, \ t \in M; \quad r,s \in R : r < s; \tag{A.29}$$

$$z_{[i,1,j][u,p,v][k,s,t]} \geq 0; \quad i, \ j, \ u, \ v, \ k, \ t \in M; \quad p,s \in R : 1 < p < s. \tag{A.30}$$

Hofman (2008) pointed out (correctly) that this relaxed model reduces to the "model without z-variables" discussed in Section 1 of this paper by adding a fictitious city to the set of cities to serve as the starting and ending point of travel. Hence, this relaxed model cannot be integral in general, as discussed in Section 1 of this appendix.

Printed in the United States
By Bookmasters